Images, Ethics, Technology

Images, Ethics, Technology explores the changing ethical implications of images and the ways they are communicated and understood.

It emphasizes how images change not only through their modes of representation, but also through our relationship to them. In order to understand images, we must understand how they are produced, communicated, and displayed.

Each of the 14 essays charts the relationship to technology as part of a complex social and cultural matrix, highlighting how these relations constrain and enable notions of responsibility with respect to images and what they represent. They demonstrate that as technology develops and changes, the images themselves change, not just with respect to content, but in the very meanings and indices they produce.

This is a collection that not only asks who speaks for the art, but also who speaks for the witnesses, the cameras, the documented, the landscape, the institutional platforms, the taboos, those wishing to be forgotten, those being seen, and the experience of viewing itself.

Images, Ethics, Technology is ideal for advanced-level students and researchers in media and communications, visual culture, and cultural studies.

Sharrona Pearl is Assistant Professor at the Annenberg School for Communication, University of Pennsylvania. Her first book, *About Faces: Physiognomy in Nineteenth-Century Britain*, was published in 2010. She is currently working on a book entitled *Face/On: Face Transplants and the Ethics of the Other.*

Shaping Inquiry in Culture, Communication and Media Studies
Series Editor: Barbie Zelizer

Dedicated to bringing to the foreground the central impulses by which we engage in inquiry, the *Shaping Inquiry in Culture, Communication and Media Studies* series attempts to make explicit the ways in which we craft our intellectual grasp of the world.

Explorations in Communication and History
Edited by Barbie Zelizer

The Changing Faces of Journalism
Tabloidization, Technology and Truthiness
Edited by Barbie Zelizer

The Politics of Reality Television
Global Perspectives
Edited by Marwan M. Kraidy and Katherine Sender

Making the University Matter
Edited by Barbie Zelizer

Communication Matters
Materialist Approaches to Media, Mobility and Networks
Edited by Jeremy Packer and Stephen B. Crofts Wiley

Communication and Power in the Global Era
Orders and Borders
Edited by Marwan M. Kraidy

Boundaries of Journalism
Professionalism, Practices and Participation
Edited by Matt Carlson and Seth C. Lewis

Images, Ethics, Technology
Edited by Sharrona Pearl

Images, Ethics, Technology

Edited by Sharrona Pearl

Routledge
Taylor & Francis Group
LONDON AND NEW YORK

First published 2016

by Routledge
2 Park Square, Milton Park, Abingdon, Oxon OX14 4RN

and by Routledge
711 Third Avenue, New York, NY 10017

Routledge is an imprint of the Taylor & Francis Group, an informa business

British Library Cataloguing-in-Publication Data
A catalogue record for this book is available from the British Library

Library of Congress Cataloging-in-Publication Data
Images, ethics, technology / edited by Sharrona Pearl.
pages cm. – (Shaping inquiry in culture, communication and media studies)
Includes bibliographical references and index.
1. Visual communication. 2. Communication and technology.
3. Technology–Moral and ethical aspects. I. Pearl, Sharrona, editor.
P93.5.I4669 2015
302.2′2–dc23
2015015851

ISBN: 978-1-138-93513-6 (hbk)
ISBN: 978-1-138-93515-0 (pbk)
ISBN: 978-1-315-67759-0 (ebk)

Typeset in Sabon
by Swales & Willis Ltd, Exeter, Devon, UK

Contents

PART II
Memorializing images

PART III
Embodying images

Figures

Contributors

Lisa Cartwright is Professor of Communication and Science Studies at the University of California, San Diego, where she also teaches in critical gender studies. Her research interests span visual studies; gender and sexuality studies; science, technology, information and medicine studies; and disability studies. She is the author of *Moral Spectatorship: Technologies of Voice and Affect in Postwar Representations of the Child* (Duke University Press, 2008) and *Screening the Body: Tracing Medicine's Visual Culture* (University of Minnesota Press, 1995). She is co-author, with Marita Sturken, of *Practices of Looking: An Introduction to Visual Culture* (Oxford University Press, second edition 2008). With Paula Treichler and Constance Penley, she co-edited *The Visible Woman: Imaging Technologies, Gender and Science* (New York University Press, 1998).

Roderick Coover is Associate Professor of Film and Media Arts at Temple University, where he teaches courses in visual research, experimental media arts, and cinema. A recipient of Mellon, LEF, Whiting and Fulbright awards, his recent works include *Switching Codes: Thinking Through Digital Technology in the Humanities and Arts* (University of Chicago Press, 2011), *Unknown Territories* (http://unknownterritories.com), *From Verite to Virtual* (Documentary Educational Resources, 2007), and *The Theory of Time Here* (Video Data Bank, 2007). He is currently at work on a multi-monitor installation and book project about Philadelphia and the Delaware River in an age of global warming.

Nora Draper is Assistant Professor of Communication at the University of New Hampshire. Her research examines the complexities of authenticity, privacy, identity, and reputation in the digital era through frames of critical cultural theory and policy. Her work has been published in the *International Journal of Communication*, *Critical Studies in Media Communication*, *Journal of Children and Media*, and *Surveillance & Society*.

Kelly Gates is Associate Professor of Communication, Science Studies, and Critical Gender Studies at the University of California, San Diego. Her research examines the social and political dimensions of computerization and surveillance system development in post-war United States. Her book, *Our Biometric Future: Facial Recognition Technology and the Culture of Surveillance* (New York University Press, 2011), examines the effort underway since the 1960s among program computers to "see" the human face. She is also the editor of *International Encyclopedia of Media Studies, Vol. VI: Media Studies Futures* (Wiley-Blackwell, 2013).

Nicholas Gilewicz is a doctoral student at the Annenberg School for Communication at the University of Pennsylvania. He researches the history of journalism, how journalists construct the social meaning of their work through journalism products, and how to theorize new frameworks with which to analyze texts and communities produced by digital communication, some of which is forthcoming in *Journalism*. He has received top-paper awards from the Association for Education in Journalism and the International Communication Association.

Kevin Gotkin is a PhD candidate at the Annenberg School for Communication at the University of Pennsylvania. His work looks at the relationships between disability, technology, and mediation. He is interested in amateur medical knowledges, visual and digital media-making practices, and the histories of science and technology.

Larry Gross is Director of and Professor at the University of Southern California Annenberg School for Communication. His research interests include the nature of symbolic communication; art and communication; media and culture; and sexual minorities and the media. He is author of *Contested Closets: The Politics and Ethics of Outing* (University of Minnesota Press, 1993) and *Up from Invisibility: Lesbians, Gay Men and the Media in America* (Columbia University Press, 2001). He is editor or co-editor of *Communications Technology and Social Policy* (Wiley, 1973), *Studying Visual Communication* (University of Pennsylvania Press, 1981), *Image Ethics: The Moral Rights of Subjects in Photography, Film And Television* (Oxford University Press, 1988), *On the Margins of Art Worlds* (Westview, 1995), and *The Columbia Reader on Lesbians and Gay Men in Media, Society, and Politics* (Columbia University Press, 1999). Co-founding editor of the online-only *International Journal of Communication*, Gross is an elected fellow of the International Communication Association and its 2011–2012 president.

Orit Halpern is Assistant Professor of History at the New School for Social Research and Lang College, and an affiliate in the Design MA program at the School of Art and Design History and Theory at Parsons. Her research is on histories of digital media, cybernetics, art, and design. Her most recent book project, *Beautiful Data* (Duke University Press), is a history of interactivity and visualization in cybernetics and design. Halpern's published works and multi-media projects have (or will) appear in *C-theory*, *Configurations*, *Post-Modern Culture*, *Interdisciplinary Science Reviews*, *Journal of Visual Culture*, and at ZKM in Karlsruhe, Germany.

Sharrona Pearl is Assistant Professor of Communication at the Annenberg School for Communication at the University of Pennsylvania, where she is also core faculty in the Program on Gender, Sexuality, and Women's Studies. Her research focuses on faces and the role of visual judgment in building relationships, which she examined in her first book *About Faces: Physiognomy in Nineteenth-Century Britain* (Harvard University Press, 2010). She is currently exploring face transplant surgery as a way to analyze the social and cultural stakes for facial manipulation and interrelationality.

Constance Penley is Professor of Film and Media Studies and Co-Director of the Carsey-Wolf Center at the University of California, Santa Barbara. She is a founding editor of *Camera Obscura: Feminism, Media, Cultural Studies* and the author of *The Future of an Illusion: Film, Feminism, and Psychoanalysis* (University of Minnesota Press, 1989) and *NASA/TREK: Popular Science and Sex in America* (Verso, 1997). She is editor or co-editor of many books including *Male Trouble* (University of Minnesota Press, 1992), *The Visible Woman: Imaging Technologies, Gender, and Science* (New York University Press, 1998), *The Analysis of Film* by Raymond Bellour (Indiana University Press, 2000), and the forthcoming *The Feminist Porn Book: The Politics of Producing Pleasure*. She has collaborated on the large public art projects *Primetime Contemporary Art: Art by the GALA Committee as Seen on Melrose Place* and *Biospheria: An Environmental Opera*. Penley was a 2009 winner of a MacArthur Digital Media and Learning Grant for *DigitalOcean: Sampling the Sea*.

Amit Pinchevski is Senior Lecturer in the Department of Communications and Journalism at The Hebrew University of Jerusalem, Israel. He is the author of *By Way of Interruption: Levinas and the Ethics of Communication* (Duquesne University Press, 2005) and co-editor (with Paul Frosh) of *Media Witnessing: Testimony in the Age of Mass Communication* (Palgrave Macmillan, 2009/2011) and (with Nick Couldry and Mirca Madianou) of *Ethics of Media* (Palgrave Macmillan, 2013).

Carrie A. Rentschler is Director of the Institute for Gender, Sexuality, and Feminist Studies and Associate Professor and William Dawson Scholar of Feminist Media Studies in the Department of Art History and Communication Studies at McGill University. She is the author of *Second Wounds: Victims' Rights and the Media in the U.S.* (Duke University Press, 2011) and is currently writing a book on the cultural legacies of the 1964 Kitty Genovese murder in New York City. Her other research examines mass-mediated representations of suffering and models of citizenship, the gender politics of environmental security and its publicity, the media activism of social movements, women's self-defense as a form of feminist pedagogy, and the gendered politics of fear.

Steven Rubin is Associate Professor of Art in the Photography program at Pennsylvania State University. Previously, he worked around the world for more than 20 years as a photojournalist and documentary photographer. A 2013 Fulbright-Nehru Scholar in India, he was also a Nieman Fellow at Harvard, an Alicia Patterson Journalism Fellow, and a Community Fellow and Media Fellow with the Open Society Institute. His current projects investigate the rise of wind energy in Kansas, the effects of Marcellus Shale gas drilling in Pennsylvania, and the precarious conditions of Burmese Chin refugees in Mizoram State, India.

Alexandra Sastre is a PhD candidate at the University of Pennsylvania's Annenberg School for Communication. Her research examines the body as a critical communicative tool, addressing how its representation within and outside the digital sphere produces particular subjectivities. Her interdisciplinary work unites communication with queer and feminist theory, critical race theory, postcolonial theory, and Latina/o studies, and has been published in the journals *Celebrity Studies*, *Feminist Media Studies*, and *Communication, Culture and Critique*.

Marita Sturken is Professor and former Chair in the Department of Media, Culture, and Communication at New York University, and a former professor at the Annenberg School for Communication at the University of Southern California. She is author of *Tangled Memories: The Vietnam War, the AIDS Epidemic, and the Politics of Remembering* (University of California Press, 1997) and (with Lisa Cartwright) *Practices of Looking: An Introduction to Visual Culture* (Oxford University Press, second edition 2009). Her most recent book is *Tourists of History: Memory, Kitsch, and Consumerism from Oklahoma City to Ground Zero* (Duke University Press, 2007).

Ira Wagman is Associate Professor of Communication Studies in the School of Journalism and Communication at Carleton University in Ottawa, Canada. He has published widely in the areas of the history of communication, media policy, and communication theory. His current research deals with early television history and the relationship between communication and time.

Introduction

Relating images

Sharrona Pearl

The changing forms of technology implicate images in new ways, but it is the images themselves that are subject to change. As this volume as a whole makes clear, images change not only through their modes of representation, but also through our relationship to them. That constant in the wildly and playfully different stories of images, ethics, and technology drives this collection; in order to understand images, we must understand our relationship to them and how they are produced, communicated, and displayed. Each of these chapters charts our relationship to technology as part of a larger complex social and cultural matrix, highlighting how these relations constrain and enable notions of responsibility with respect to images and what they represent. To put it another way, as technology develops and changes, the images themselves change, not just with respect to content, but also in the very meanings and indices they produce.

Images are, and must be, conceptualized as relational. They are meaningless if unseen, unexperienced. And that is why "ethics" figures prominently in the title to this collection; though we may think of the image as the object of analysis, and the technology as the medium of communication, the ethic is the relationship between these two and those who view them. All are constantly in flux. It is the work of these chapters to chart the location and connection between them at particular moments in time. In so doing, they highlight the importance of reflexivity in our conceptualization of what constitutes the image and what complex of relations it enacts by its very being or, rather, by its very seeing: by the very act of its being seen, and by those who produce it. There is a labor story here, as all these chapters show, and it concerns too those for whom images are produced.

As these chapters move through various moments in history (many of which are anchored in the nineteenth century, when technologies of vision were undergoing a radical reconfiguration that is in many ways resonant with today), they highlight that change is a constant. So the volume's claim to timeliness – to assembling the juxtaposition of images, ethics, and technology in this particular historical moment – is not one of specificity of any particular media, though that certainly ramifies in the changing configuration of what it is possible to look at. The ramifications of this reflection are left to the authors themselves, who each make the very clear case that there are ethical stakes to images and their communication, and that there are ethical stakes to the ensuing moments of reconceptualization. This is a particularly resonant question *right now*, as the increasing digitization of that which we look at may lead us to believe that our relationship to images can be standardized. By unearthing the political, social, and cultural stakes around what images mean and how that changes, this volume underscores not just *why*, but also *how* the relational nature of the image experience is personal, contingent, contextual, and deeply, deeply human.

This is an intervention. These chapters, drawn from the December 2012 symposium of the Scholars Program in Culture and Communication at the Annenberg School for Communication at the University of Pennsylvania, move away from a strict Barthesian approach of visual semiotics and image-as-text. When they read images, they do so in a larger conversation with context and practice, with careful attention paid to the reciprocal relationship between image and viewer. Equally, they depart from the classic critiques of image-as-text, as found in Nelson Goodman, Ernst Gombrich, Gotthold Ephraim Lessing, and Edmund Burke, discussed in depth by W.J.T. Mitchell.[1] They don't concentrate on listing the differences between, say, paintings and poems. Rather, these chapters activate images as objects of engagement, both theoretical and practical, as a means of examining the relationship between images and viewers, images and contexts, and images and the ethics they invoke.

Communication and images

From its early inception as a disciplinary category, communication has participated in the discussion around representation through its particular engagement with audiences, relationships, and questions of response. Images have always been an important part of communication and its academic and scholarly pursuit and organization. Communication, at its core, deals with how people understand and relate to one another, and how these ideas and understandings are disseminated on both small and large scales. Communication, at its core, understands that these ideas are always mediated, and that these mediations are often highly visual. Communication, at its core, has always already been about images, ways of seeing, and ways of being seen. The communication discourse around images reflects a grounded engagement with theory, practice, and the relationship between the two. The field has drawn on its legacies in both quantitative and qualitative methods to place images in a broader context of use. The chapters in this volume take the best of these traditions to focus our attention to how images are used, what we do with them, and what they do to us. While these discussions are anchored in images, the questions the authors ask are ones of relationships between images, technology, and their ethical implications and enactments. At their core, they are questions of communication.

W.J.T. Mitchell has helpfully distinguished between visual studies and visual culture by noting that the former is the study of the latter, an approach that "avoids the ambiguity that plagues subjects like history, in which the field and the things covered bear the same name." He continues to note that "we often confuse the two" and so he "prefer[s] to let visual culture stand for both the field and its content."[2] Mitchell's piece directly engages with the challenges inherent in defining and critiquing any emergent and/or interdisciplinary field: we don't know what to call it, where it starts, where it finishes, who specifically does it, and whether we ought to be bounding it at all. This volume is promiscuous in its methodology and disciplinary commitments precisely in order to let the images and the practices around them speak – if not for themselves, then with their interlocutors and through the work they do. Rather than let disciplinary divides dictate the conversation, the chapters turn to the images to dictate their methodologies as a means to explore what are questions of relationships and questions of communication. In order to understand visual culture and the work of images, we must understand how they are produced, disseminated, and communicated. We follow Mitchell's lead in our conceptualization of this volume as fundamentally one of communication in both field and content. In so doing, we stake a claim that the study of images is the study of communication, as it has always been.

Early communication scholars were deeply engaged in the work images do, paying attention to the effects of early film, and, later, television, including now highly controversial studies from Herbert Blumer and Philip Hauser as part of the Payne Fund Studies, Carl Hovland's studies of persuasion and attitude change, and the well-known People's Choice Study conducted by Paul Lazarsfeld, Bernard Berelson, and Hazel Gaudet.[3] These largely quantitative and survey-based projects placed significant stock in audience reaction and response, offering a number of methodological innovations around image and visual analysis in the context of viewership and audience, as did later works in the "Uses and Gratifications" model, all which thought not only about images but also about what they do and what is done with them. These studies were not framed in terms of images and often made no distinction between image, text, and sound, and did not apply specific methodological attention to the visual. Nevertheless, images remained at the center of their focus and attention, such that these works continue to be interesting references (that act almost as primary sources) when considering the history of the field and its approach to visuality and images.

The more qualitative approaches of cultural studies of communication have explicitly engaged the language of visual studies and image analysis in their approaches to communication. Scholars have long turned to the theories of Walter Benjamin and the associated Frankfurt School to think critically about the role that the circulating of images – both moving and still – plays in broad political and cultural attitudes. The approach of these scholars was highly cynical about the mass media and its cultural power, paying little or no attention to the possibility of audience autonomy and decision-making. These gaps set the stage for later cultural studies of audience response, most notably approaches that came from David Morley, Stuart Hall, and the Centre for Contemporary Cultural Studies (Birmingham, UK) during the 1970s, which drew on semiotics in media texts. Such works, particularly the Encoding/Decoding model, are both direct engagements with the Frankfurt School and important methodological models for the ways that the varieties of audience responses can be studied on both the individual and the collective levels.[4] Hall and others, including Nicholas Abercrombie and Virginia Nightingale, are particularly concerned with television and film, making the important and controversial move to frame visuals as readable, and using textual techniques to reignite interest in the image while never losing sight of the relationship between images and viewers.[5]

Later communication scholars such as Jon Cruz, Justin Lewis, Andrew Hart, and Tim O'Sullivan have usefully expanded their approach to encompass print advertising, public interest announcements, billboards, and even live-action performances.[6] They consider both media and audiences to produce a fuller engagement with the image experience. Communication theorists in particular have spent a lot of time thinking about audiences and how they work with respect to the texts they read and consume. As a few innovative scholars outside communication have long grasped, these methodologies and studies have enormous relevance to visual culture and how it is understood. Historian and literary scholar Michael Wolff recounts a visit in 1968 to the Centre for Contemporary Cultural Studies to discuss "applying their innovations to the nineteenth century" in order to breathe new life into archival materials and to understand historical audience reactions in new ways.[7] The key innovations Wolff applies are those engaged with reciprocity between images and those using and consuming them. Communication scholars have always understood that as we do things with images, images both cause change and are subject to it.

The study of images as relationships – inspiring them, catalyzing them, disrupting them, scripting them – has been squarely in the domain of communication scholarship. Communication scholars have always considered images in contexts and as relationships: between maker and viewer; between object and audience; between material culture and cultures of production; between history and the present; between memory and possibility. In our relationship with images, there are competing voices, competing claims, and competing goals. Images have and confer authority, they create and archive memory, and they represent and construct bodies. And they sometimes sow discord and disagreement; as these chapters also show us, we don't always agree what to do with images.

Following in the communication tradition, all of the scholars in this volume are concerned with images, technology, and the relationships they engender. When we do things with images, we invite responses and reactions, and these have ethical stakes that must be explicitly discussed. Ethics are an important and oft-neglected part of the image conversation, with some exceptions; in 2000, Larry Gross, John Stuart Katz, and Jay Ruby convened a conference entitled "Image Ethics in the Digital Age" to consider what particular role the digital played in the ethical terrain around images. The resulting volume asks a question: what kinds of new social relations are produced by new technologies, and how do they undermine or challenge traditional ways of doing things with images? What evaluative criteria ought we to apply to how we use images in this new relational context?[8]

Images, Ethics, Technology, while interested in Gross et al.'s questions, has a different focus; rather than asking how we ought to evaluate our use of images, it takes a step back and wonders whether we do so at all, or, rather, *who* does so. This question puts the ethical stakes at an earlier stage, considering labor and production practices; it thinks about the way these issues frame use, rather than focusing on use itself. What is the voice of authority, and where is it situated? Who, as Larry Gross provocatively asks here, speaks for the art? Or the witnesses (Rentschler) or the cameras (Gates) or the documented (Coover) or the landscape (Cartwright and Rubin) or the institutional platforms (Sturken) or the taboos (Penley) or those wishing to be forgotten (Wagman) or those being seen (Pinchevski) or the experience of viewing itself (Halpern)?

This volume suggests that the process of reflection on changes in that which we see – be it the rural landscape of Kansas, as Lisa Cartwright and Steven Rubin chronicle; the computational and visual technology that Kelly Gates describes; or the reconfiguration of witnessing and bystanding that Carrie A. Rentschler proposes – is itself an ethical process. It is a political, sociological, and philosophical one too, according to the theoretical guide that Amit Pinchevski offers. The visual is an index of experience, a manipulable one, as Roderick Coover documents, and by its very nature must have a politic. There are ethical stakes to understanding that the contents of images are subject to their context; as Constance Penley teaches us, that which counts as obscene is subject to who is looking, and where, and under what laws and political motivations. As these laws and politics and contexts change, the framework by which we understand looking itself changes, and sometimes, as Ira Wagman suggests, the technology itself needs to be reconfigured according to ethical imperatives. The law too, Larry Gross highlights, is subject to context. And while perception itself may be cast as inhuman, as Orit Halpern charts, the experience of the visual is very human indeed. And it is that humanness to which each of these chapters turn, a particularly resonant reminder of the corporeal, the visceral, and the somatic in this moment when our attention is being ever more focused on the mechanical, the digital, and the automatic.

Outline of the book

The connection between images, ethics, and technology is a relational and contextual one. It's an activating one: we do things with images, and images make us do things. This demands an engagement on the part of people, the humans, who make meaning of it all. To underscore the nature of these relationships, this volume is divided into three thematic parts – authorizing, memorializing, and embodying – each of which is an action in its own right, as well as an argument about the work that images do in the world, and they ways in which we react to them. It asks: what is the source of the authorizing, the memorializing, and the embodying work that we do with images, and that images do in the world? These sections are organized around practices, because images, as these chapters all remind us, both act and enact. They authorize, memorialize, and embody. At the same time, we authorize images, memorialize them, and embody them. And, in so doing, they cause us to reconsider the nature of authority, the relationship between memory and history, the experience of the body, and the networks around it. There's another relationship that is central to this volume, between the images and those who view them. All of these chapters operate with a keen awareness of the real-life stakes for their objects and investigations, and they all pay particular attention to questions of management, negotiation, and response, as images demand that they must.

Authorizing images, for the purposes of this volume, are those used to give permission and those given permission to confer power in legal and evidentiary contexts. The stakes here are no less than the framework for democratic discourse. Institutional authority, including that vested in the law, is a construct developed for the maintenance of social order and public safety to provide a mechanism to achieve consensus, settle disputes, and avoid violence as a way to resolve conflict. No one image, and indeed nothing, has immutable authority; it's a process that involves the conferral of power and its visual representation. All of the discussions in this part highlight questions of responsibility: that which is invoked by the image, that which is required toward the image, that which is undermined by the contexts in which the image exists. The chapters by Carrie A. Rentschler, Kelly Gates, Constance Penley, and Larry Gross all consider the process by which images are vested with authority and how they exert it, and the debates around these mechanisms. As Nora Draper highlights, they "raise important questions about the ethics of image possession, the cultures of distribution and circulation, and the nature of mediated witnessing in the digital era." They each work to locate perceived authority in different spaces, which "queries the stakes for contemporary practices of observation and the challenging role of the mediated image as a site of authority."

Memorializing images are those that do the work of ordering our history and framing the way we represent our individual and collective pasts. But the past is also contested terrain, not only in terms of what happened, but also who gets to describe it. There is a fantasy of historical transparency in which images participate, providing as they do a documentary function. But our illustrated stories of the past are (usually) carefully curated and collected, a process that is itself highly untransparent. This volume is invested in uncovering the decisions and accidents that contribute to our memorializing images, and thinking through the contemporary stakes for the way that we use images to memorialize our lives and the lives of others. Sometimes we wish to erase the images that stand as a prompt and proxy for memory. Who gets to decide on the inclusion and erasure of images in media and in our private lives? What are the hidden costs of image-sharing? To whom does memory belong?

These questions are addressed by Marita Sturken, Ira Wagman, and Roderick Coover in the opening section. They are hard to isolate, as Kevin Gotkin reminds us, since "memory is hard to hold, slippery when we try to contain it." It is the work of these chapters to "reckon with the wily nature of contemporary images . . . and in so doing, [they] model a capacious framework for understanding the intersection of memory, history, and visuality." What emerges from this part is the ephemeral nature of memory, and the ways that images can serve as a site to discuss and concretize not just the debates about memorializing, but also memory itself.

Images are things, but they also capture and represent things, and in so doing tell stories and encapsulate experiences. Images embody, and images are of the body. And images become embodied both in the sense of representing a particular set of ideas, and in terms of becoming alive, thereby creating sensory engagement and making us feel differently in our own bodies. Images then manipulate, and are manipulated to change the way we interact with one another, and to change the way we see one another. This part, with chapters by Orit Halpern, Lisa Cartwright, Steven Rubin, and Amit Pinchevski, highlights the corporeality of viewing, the experiential nature of interacting with images, and the stakes for the metonymical power of an image, which is part of something larger, to represent that whole. They do so by thinking about images of events, landscapes, and even humanity itself by emphasizing, in the words of Alexandra Sastre and Nicholas Gilewicz, "the oft-elided processes of subjectification that can govern the creations of images and our understandings of perception." It matters, they emphasize, because "the power of images lies precisely in their ability to produce subjects . . . and that subjectification informs how we see the bodies and perceive the minds of those people."

All three parts make clear the ways that images are both concrete – as objects – and abstract – as processes themselves and archives of their own production. It is this combination of materiality and ephemerality that causes images to be such powerful makers of meaning and agents of action. They do things by their very being, and through the histories and presents that their being encapsulates and represents. The chapters all consider questions of iconicity, wondering what we are allowed to change and showing that not only images and their subjects become iconic, but also their platforms, their media, and their technological frameworks. They remind us that there are human actors and human relationships at play in the production of images, in the institutional housing of these images, in the economic affordance and structures around these images, and in the images themselves. There is careful attention paid to the work of the images, both in how they function in the world and, equally if not more importantly, how they come to be. Creating a past, making memory, making history is work. It is deeply situated, and highly charged labor, and it is with these deeply rooted contextual and relational considerations that we can consider its ethical implications. These chapters all pay careful attention not only to the content and context of images, but also the interaction – necessarily a relational and an ethical one – between the images and those who view them, and the changes this process itself brings into being.

Notes

1 W.J.T. Mitchell, *Iconology: Image, Text, Ideology* (Chicago: University of Chicago Press, 1986).

2 W.J.T. Mitchell, "Showing Seeing: A Critique of Visual Culture," *Journal of Visual Culture* 1 (2002): 165–181.

3 The Payne Fund Studies included eight volumes published between 1933 and 1935. A sample study can be seen in Herbert Blumer and Philip Hauser, *Movies, Delinquency, and Crime* (New York: Macmillan, 1933). Carl I. Hovland, Irving L. Janis, and Harold H. Kelley, *Communication and Persuasion: Psychological Studies of Opinion Change* (New Haven: Yale University Press, 1953). Paul Felix Lazarsfeld, Bernard Berelson, and Hazel Gaudet, *The People's Choice: How the Voter Makes up His Mind in a Presidential Campaign* (New York: Duell, 1944).

4 Stuart Hall, *Culture Media Language: Working Paper in Cultural Studies, 1972–79* (London: Hutchinson/Centre for Contemporary Cultural Studies, University of Birmingham, 1980).

5 Nicholas Abercrombie, *Television and Society* (Cambridge, Mass.: Blackwell, 1996). Virginia Nightingale, *Studying Audiences: The Shock of the Real* (London; New York: Routledge, 1996).

6 Jon Cruz and Justin Lewis, *Viewing, Reading, Listening: Audiences and Cultural Reception* (Boulder, Col.: Westview Press, 1994). Andrew Hart, *Understanding the Media: A Practical Guide* (London; New York: Routledge, 1991). Tim O'Sullivan, Brian Dutton, and Philip Rayner, *Studying the Media: An Introduction* (London; New York: Routledge, Chapman, and Hall, 1994).

7 Miles Taylor and Michael Wolff, *The Victorians since 1901: Histories, Representations, and Revisions*, (Manchester, England: Manchester University Press, 2004).

8 Larry Gross, John Stuart Katz, Jay Ruby, eds., *Image Ethics in the Digital Age* (Minneapolis: University of Minnesota Press, 2003).

Part I

Authorizing images

Chapter 1

Introduction

Interrogating the authority of the image

Nora Draper

In August 2014, a poignant question about the nature of image ownership arose across social media outlets: If a monkey takes a selfie, who owns the copyright? The question was not hypothetical. In 2011, British photographer David Slater had traveled to Indonesia to take pictures of the local wildlife. During the course of his trip, a crested black macaque stole his equipment and took hundreds of pictures of itself. The images included a particularly beautiful shot that was reproduced and entered by the Wikimedia Foundation into a bank of public domain images to be used free of charge. Slater, who often self-finances his photography trips, has argued that the image is not in the public domain. He contends that he owns the image copyright because the photo was taken with his camera during the course of his work as a nature photographer. Wikimedia – and a number of copyright experts – disagree.[1]

This debate is not trivial – nor do its implications reside solely on the peripheries of copyright law. Deliberations about the nature of ownership and use of digital images raise important questions about the ethics of image possession, the cultures of distribution and circulation, and the nature of mediated witnessing in the digital era. We tend to be comfortable with the idea that an image's authority is derived from the intentions and credibility of its producer. In a digital moment, however, the experience of mechanically produced pictures, recombinant images, and suspect sourcing, assumptions about the supremacy of the image have been dredged out of their comfortable, if tenuous, resting spots.

In his essay on witnessing, John Durham Peters eloquently articulates the tensions implicit in the widely used term. "Witnessing," he argues, "is an intricately tangled practice. It raises questions of truth and experience, presence and absence, death and pain, seeing and saying, and the trustworthiness of perception."[2] Peters describes witnessing as simultaneously passive (seeing) and active (giving testimony), further noting the gulf that separates the two components of the practice. This perceived void between seeing and saying reveals concerns regarding the limits of witnessing. The role of the media, which act as proxy witnesses through the translation and transmission of distant images, has often been examined for its effects on viewers.[3]

Media events theory describes how broadcast television creates conditions that allow people to understand the media's observations as their own.[4] The "liveness" of the mediated spectacle has allowed witnessing to become a domestic act.[5] Some question, however, whether such secondhand viewing allows an individual to lay claims as to having borne witness to an event. Particularly acute are concerns about the translation of emotion through images and how this affects individual reactions to misery.[6] How does an obligation of intervention that one may feel when witnessing an event in person respond to the intercession of a third-party testimony? Some academic research has offered that responses to distant suffering involve superficial charity[7] and, eventually, compassion fatigue.[8]

The unresolved status of images as authoritative sites for witnessing generates further apprehension regarding the nature of mediated observation and testimony. Debates regarding the supposed objectivity of the image and its corresponding authority have a long history. While discourse arising in the late nineteenth century privileged the objectivity of the scientific image,[9] later voices questioned these claims, noting that human intervention and cultural context are implicit in all image creation.[10] The rise of citizen journalism has led to a reexamination regarding the integration and contextualization of images from unvetted sources by news organizations.[11] Often, these debates hinge on a strained relationship between the authenticity of the amateur and the authority of the professional.

Situated within this critical intellectual tradition, each of the chapters in this section explores the role of the image as a site of authority within the context of mediated witnessing. Each chapter tackles important questions about how the rights and responsibilities of viewership are nested in an increasingly complex notion of images as evidence and challenged by gatekeepers that regulate their consumption.

The 1964 Kitty Genovese murder in New York City is familiar to us for its production of a set of anxieties around the bystander. Reports of the event noted that up to 38 witnesses watched as a young woman was stabbed repeatedly and bled to death. None of those bystanders intervened or called for help. Despite reinterpretations of the event that revised both the number of bystanders and their actions, the reporting of Kitty Genovese's murder contributed to the construction of a theory of the "bystander effect," which posits that the more people who witness something bad happening to another person, the less likely any of them will be to intervene. Applications of this theory often parallel concerns that individuals are unmoved by images of suffering. These theories suggest that, in aggregate, individuals who see a crime in progress are able to distance themselves and transfer the obligations of intervention to others.

Carrie A. Rentschler draws directly on the cultural and academic legacies of the Genovese murder to examine the construction of the bystander as a failed citizen. In noting that contemporary bystanders are often condemned for their use of camera phones to document rather than intercede, Rentschler offers an alternative framework for understanding such forms of mediated witnessing. She offers the role of bystander witnesses as a performance of active intervention and characterizes their behavior as disrupting the existing relational power of the actors.

Through an examination of the use of new visual imaging and archiving technologies in police work, Kelly Gates explores the ways in which modern investigatory and evidentiary practices have been transformed. Looking specifically at the explosion of commercial products that offer ways to enhance forensic video evidence and the new labor conditions they produce, Gates describes an emerging belief in "computational objectivity," which privileges the neutrality of technologically mediated observation over the imperfect and selective perception of the human witness. Just as Peters notes the credibility of scientific instruments to witness has been established through their "indifference to human interests,"[12] Gates observes that the promotion of forensic technologies generates expectations around authenticity and credibility at the same time as they obscure the role of human intervention in both the construction and operation of the technology and in the interpretation of its output.

In the relation of her experiences as a would-be expert witness in an obscenity trial, Constance Penley explores the implications for due process when members of a jury are asked to interpret mediated images based on their relational qualities. Of course, decisions of jury members always rest on interpretations of mediated evidence, whether it is through an eyewitness, police investigator, or CCTV footage. In the case Penley presents, however, the images are mediated in two additional ways. The first is through the decision of a judge regarding which images are viewed by the jury and under what conditions. The second is the presentation of "similar" imagery that is expected to stand in for the original pieces on the basis of perceived generic resemblance. Penley explores the implications of mediation when images are stripped of their content (and context) and miscategorized through their sameness.

In his chapter on the contentious relocation of the Barnes Foundation Art Collection from its original home in Merion, Pennsylvania to Philadelphia, Larry Gross discusses the roles and responsibilities that befall the custodians of images. The experience of any art gallery is mediated through the particular view of its curator. This is particularly true in the case of the Barnes Foundation, which houses a collection that – owing to the particular tastes of its benefactor – is at once astonishingly resonant and deeply idiosyncratic. In interrogating the debates that slowed and eventually guided the collection's move to its present location, Gross examines the simultaneous role of the collector as an authority on the value of the art in his collection and as the guardian of access. In reflecting on how the consumption of collections is mediated through the eye of the collector and the institutional forces that shape access, Gross examines how the authority conferred by ownership shapes the public experience of art.

Together, these chapters examine how questions of authority and objectivity, responsibility and ownership, are negotiated through acts of mediated witnessing. They explore how the cultural meaning of images – whether they are grainy surveillance videos, footage from a cellphone camera, highly produced pornography, or prestige art collections – is shaped when they are observed secondhand. By locating perceived authority alternatively in the expert (police, judge, or collector), amateur (bystander), or machine (digital CCTV), each author queries the stakes for contemporary practices of observation and the challenging role of the mediated image as a site of authority.

Notes

1 Jordan Weissmann, "If a Monkey Takes a Selfie, Who Owns the Copyright?," *Slate*, August 6, 2014, www.slate.com/blogs/moneybox/2014/08/06/monkey_selfie_who_owns_the_copyright.html.
2 John Durham Peters, "Witnessing," *Media, Culture & Society* 23, no. 6 (2001): 707–724.
3 Luc Boktanski, *Distant Suffering: Morality, Media and Politics*, trans. Graham Burchell (Cambridge: Cambridge University Press, 1999).
4 Daniel Dayan and Elihu Katz, *Media Events: The Live Broadcasting of History* (Cambridge, Mass.: Harvard University Press, 1992).
5 John Ellis, *Seeing Things: Television in the Age of Uncertainty* (London, I.B. Tauris, 2000).
6 Jean Seaton, "Watching the World – Seeing, Feeling – Understanding?," *The Political Quarterly* 72, no. 4 (2001): 498–502.
7 Slavoj Žiž ek, *Violence* (New York: Picador, 2008).
8 Keith Tester, *Compassion, Morality and the Media* (Buckingham, UK: Open University Press, 2001).
9 Lorraine Daston and Peter Galison, "The Image of Objectivity," *Representations* 40 (Autumn 1992): 81–128.
10 Kenneth E. Boulding, *The Image: Knowledge in Life and Society* (Ann Arbor: University of Michigan, 1961).
11 Mervi Pantti and Kari Andén-Papadopoulos, "Transparency and Trustworthiness: Strategies for Incorporating Amateur Photography into News Discourse," in *Amateur Images and Global News*, ed. Kari Andén-Papadopoulos and Mervi Pantti (Bristol: Intellect, 2011), 99–112.
12 Peters, "Witnessing ," 715.

Chapter 2

Technologies of bystanding

Learning to see like a bystander

Carrie A. Rentschler

Over the last few years, parts of Montreal's infrastructure have been giving way: hunks of concrete drop from salt-saturated highway overpasses; one overpass collapsed, dropping cars onto the freeway below; a cement slab detached from a downtown high-rise hotel and crashed into a restaurant below, killing a woman who was out to celebrate her birthday; and, in fall 2012, a sink hole the size of a large car opened up in the middle of one of downtown's busiest streets. On January 28 2013, in another sign of our city's fracturing above- and below-ground networks, a water main broke near the university campus at which I work, turning one of the streets into a surging river and flooding several buildings.

Moving through the city, Montreal residents and visitors become witness to infrastructural crisis as the new ordinary. With mobile phones close at hand, many of us record signs of the crisis, transforming our status as bystanders to structural crisis into documentarians of it. On the day of the flood, several university students uploaded cellphone videos of the floodwaters rushing down rue McTavish, providing eyewitness perspectives to the unfolding situation and inadvertently documenting their own and others' roles as proximate bystanders in the process. Audio recordings amplify the chatter that expressed the situation of bystanding that day: from nervous laughter to exclamations of surprise and concern, to assertions that someone ought to do something. Several videos record a lone university student being washed down the street in a watery torrent; she was crossing the street when the rushing waters rose over her knees, knocking her down. On the audio track that accompanies one of these videos with over 1.6 million views on YouTube six days later, listeners hear a young woman laughing; one voice exclaims that the young woman in the water should get up (Figure 2.1). A male voice quips "where is she *going*?" as if the woman could control her movements in the powerful flood waters. At the end of the recording, we hear a man declare, "Dude, somebody should probably go save her."[1]

To one journalist, the latter comment represented a single piece of evidence that bystanders were bound by a moral responsibility to help the young

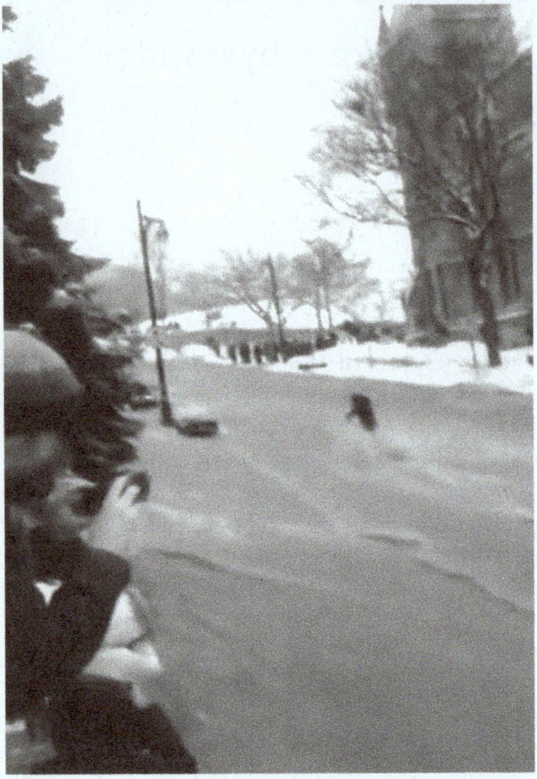

Figure 2.1 The documentary impulse of the bystander (still from a cellphone video uploaded to YouTube showing a female student at McGill University caught in a flood).

Source: www.youtube.com/watch?v=NJG3A5FXO0o&feature=youtu.be.

woman.[2] The woman eventually grabbed hold of a large planter and pulled herself out of the water, after which staff from the university bookstore provided her with dry clothes and hot cocoa. Yet as I argue in this chapter, the whole video, and not just the statement that someone ought to help, reveals the messy and contingent situation that defines bystanding in the midst of crisis situations, complicating what Kelly Oliver[3] calls the response-ability of witnesses.[4] Subjectivity, Oliver argues, is "founded on the ability to respond to, and address, others."[5] While journalists and others jumped to condemn those who recorded the young woman being washed down rue McTavish, bystanders who documented the scene and recorded others in the process of onlooking did respond to the situation, even if it was in ways that others condemn as not doing enough. The video records of their bystanding evidence that the basic conditions of the ability to respond have been met.

This chapter attempts to trouble distinctions that have been drawn between witnessing as an act of taking of responsibility and bystanding as the failure to take responsibility. I analyze bystander videos and social-movement initiatives that use mobile-phone applications to target bystanders as potential interveners in order to situate bystanding as part of the process of bearing witness to events. Today, bystander videos such as the one discussed above represent one particularly visible and mobile genre of video witnessing online, a term I revise from Paul Frosh and Amit Pinchevski's concept of "media witnessing."[6] "Video cameras," they assert, constitute "a technological surrogate for an audience of the witnessing process underway."[7] Video records of bystanding socially and technologically constitute processes of media-enabled witnessing. Due to the ubiquity of mass and network media dissemination of audio-visual recorded material, most of us are already situated as witnesses to both mundane and extraordinary events.[8] People on campus and all over Canada came to know about the flooding at McGill University because of mobile-phone recordings by bystanders and their dissemination on YouTube and other social media sites. "In the eyes of the public," archivist Rick Prelinger argues, "YouTube ha[s] become the default online moving-image archive."[9] Through ubiquitous media and online video aggregators, we may pay witness audio-visually to bystanding, but we do not necessarily *bear* witness as a form of responsibility for what we see and hear.

Following from Sue Tait's[10] work on the concept of bearing witness, I examine bystander videos for how they distinguish between the eye- and ear-witnessing of an event, and *bearing witness* to it as an activity of taking responsibility.[11] In the act of recording bystanding, depicting it, and acting as cultural technical bystanders themselves, bystander videos dramatize the contingencies, insecurities and fallibility of bystander response-ability. In many cases, the bystander experience may ultimately fail to deliver on this capacity to respond in an interventionist way, but I argue that it ought not to be interpreted as a failure to care or be moved affectively in the right or "correct" ways, or used as proof of an absence of the capacity to act itself. Rather than excuse bystander non-intervention, my interest lies in understanding how bystanding has become the site of potential social intervention for activists and social movements, as they target individuals' abilities to respond as proximate participants, often in modes that are further enabled by mobile phones. Today the limitations of acting as a bystander are also now the grounds for activist potential.

For some advocacy organizations such as the human-rights group WITNESS, video provides a link "in between the extremes of undifferentiated mass media attention and direct evidence in the courts . . . a 'space for action' by the audience."[12] WITNESS uses video to move audiences from the position of those who see to those who have the capacity to act: from eye- and ear-witness to people who can bear witness in the act of testifying.[13]

For Peter Walker, Director of the Feinstein International Center at Tufts University, "Cheap and ubiquitous access to video cameras on cell phones, combined with the melding of cell phone networks and the Internet, will allow those caught up in human rights atrocities to tell their stories directly to the world. States will find it increasingly difficult to conceal this resistance."[14] From the perspective of WITNESS, video does not provide neutral evidence of human-rights abuses; it offers testimony, ways of bearing video witness within contexts of meaning making and dissemination directed toward the promotion of action.[15] If such practices of media witnessing put "society on view to itself," videos produced by bystanders deploy an additional audio-visual logic, putting the act of bystanding and the position of the bystander under view, and in some cases mobilizing bystanders as technologically enabled interveners.[16]

This chapter analyzes how activist and non-activist video making stages the problem of bystander non-intervention in ways that re-define it as a position from which one could bear witness. As the flooding case and many others suggest, video documentation serves as a key technology of bystanding, where the fact of being in proximity to events with mobile media technologies that are close-to-hand enables subjects to document them. I examine how advocates deploy networked video-production technologies in the process of transforming bystanders into interveners. After defining my approach to bystanding, the chapter proceeds with an examination of how social-movement organizations and educational initiatives define the bystander as a target of social intervention, as subjects positioned to intervene in violence and harassment committed against others based not only on their proximity to scenes of social violence, but also as members of the communities in which the violence occurs. The chapter then turns to analysis of how the online aggregation and dissemination of bystander videos makes bystanding increasingly audio-visible in forms that stage it as a contingent, transformable practice.

A movement-identified approach to bystanding

Social movements currently refigure the position of the bystander, not as a failed intervener but as someone with the capacity to intervene. From anti-bullying initiatives in schools and human-rights education to feminist activism against street harassment and domestic violence, social movements target the bystander as a site of social intervention, and they use networked mobile technologies in order to mobilize an activist-bystander orientation to problems of social violence. Through their work, I approach bystanding as a practice constituted through the video documentation of proximate events. While social definitions of the bystander frame him or her as a non-intervener, someone who sees or witnesses an event but does not intervene, movements and educators increasingly define bystanders as people positioned to witness events and intervene in them using the tools of mobile-phone video, audio recording and Google mapping techniques.

According to Emily May, executive director of the international anti-street harassment organization Hollaback!: "There is a really important role for bystanders to play. I think too often people think they only have two options: either swoop in wearing a super hero costume and beat the guy up, or do nothing."[17] For May, bystanders can intervene in their role as documentarians of street harassment. Hollaback! trains individuals and community organizers how to use iPhone and android phone applications to document, map and narrate acts of street harassment they have been targets of or have directly witnessed happening to others (Figure 2.2). They model activist-bystander intervention using mobile-phone technologies, teaching women and girls to take the tools of mobile media documentation into their own hands. Through Hollaback!'s activism, the cellphone becomes a networked video-production and documentation tool against street harassment, an activist-bystander technology.

From a movement perspective, bystanding signifies a capacity for action in ways that academic scholarship tends to identify with the more active role of witness testimonial. "To witness an event," as John Peters argues, "is to be in some way responsible to it"[18]: to move from seeing to saying. For Peters, the act of seeing is passive and often accidental, while speaking represents active and purposeful activity. The latter marks the witness, a *person* and not a technology, as a "privileged possessor and producer of knowledge."[19] While a witness can be a person, a text, a media artifact, a technology or an audience, the human subject figures centrally in theories of witnessing and, by extension, conceptions of bystanding. As an act, witnessing constitutes a sensory experience of hearing, seeing, touching and smelling an event in process; it also constitutes an act of translation between sensory experiences and the statement or representation of them to others, as testimony.

Activists approach bystanding – the position of being proximate witness to social violence and cultures harassment – as a *capacity for action*. They see the bystander as a participant-witness in social violence and cultures of harassment situated in-between the position of perpetrator and victim. In the second of a series of three video blogs produced by the local Winnipeg, Manitoba Hollaback! site, creators Jodie Layne and Cleo Leslie describe the bystander as someone with the power to intervene in a harasser's harassment of their victim. Layne asks viewers "what would you do to be an *effective* bystander?"[20] suggesting that bystanding is not a failure to act but that it signifies instead the capacity to act more or less effectively. Their third video blog delineates different kinds of effective bystander behavior such as making eye contact, pointing out the harasser to others, and creating verbal and physical barriers between harassers and victims.[21] Their models of bystander behavior situate the bystander as a link to a larger community of participants

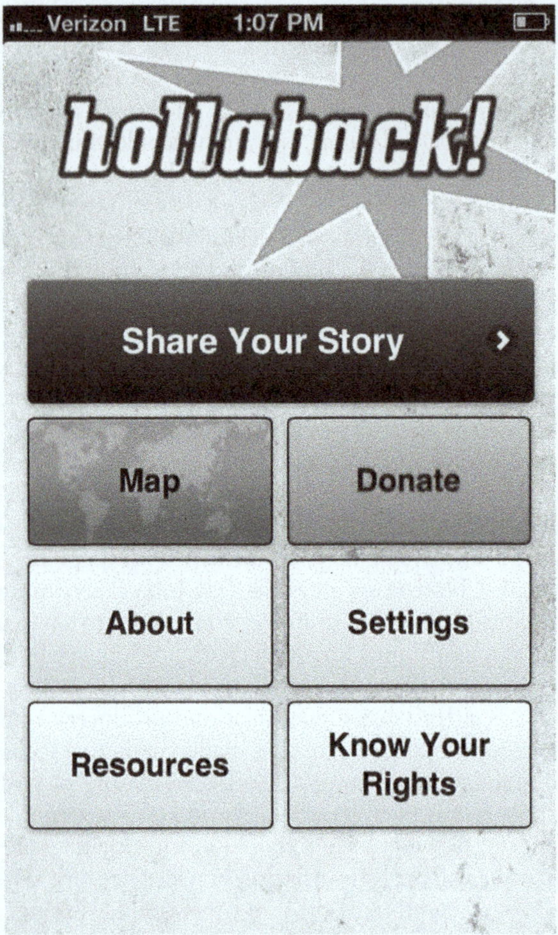

Figure 2.2 Hollaback!'s iPhone application.

and potential interveners. Breaking out of the criminological dyad of perpetrator and victim,[22] the bystander appears as a third position linking the position of witness to violence to the role of the community in responding. The bystander position thus stands in for the larger community of witnesses to the perpetration and experience of violence in context (Figure 2.3).

While the bystander role is typically described as either being an accomplice to violence (as a bystander-participant) or an innocent victim of circumstance (that of the innocent bystander), the quality of being in-between or alongside that of victim or perpetrator as someone in proximity to the scene of intervention creates an opportunity to re-conceptualize the problem of the

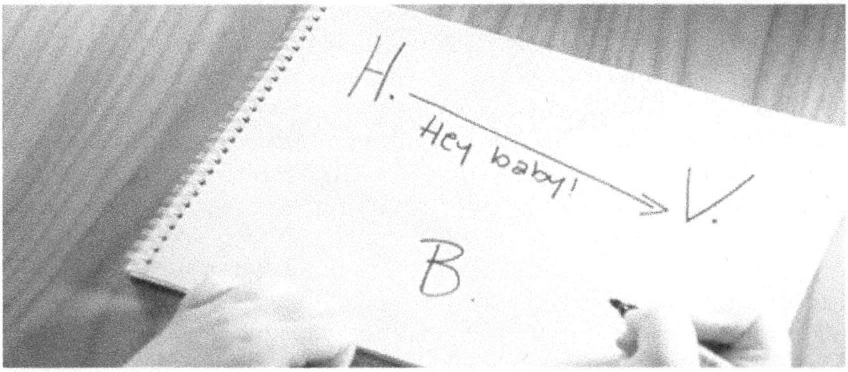

Figure 2.3 The bystander as a position of witness between harassers and their victims (still from Hollaback! Winnipeg video blog "An Introduction to the Bystander").

Source: www.youtube.com/watch?v=HlrGxZCy9Hw.

bystander from the perspective of those who seek to redefine it as a site of possible agency.[23] Situated between victims and perpetrators, bystanders can enable harassment and violence or they can disrupt it. For Hollaback! and other groups, bystanders' failures to intervene signify not a lack of action so much as another form of participation in harassing situations. Whether bystanders fear reprisal or getting hurt, support the harassment, feel helpless to stop it, feel judged or under scrutiny from others, or do not know what to do to intervene, the bystander position is always already one of participation that comes with consequences.

Hollaback! seeks to transform the act of bystanding into interventionist behavior by first teaching people to see bystanders differently: as interveners rather than failed subjects. Following Naomi Mandel's[24] reconceptualization of complicity as a form of participation prior to judgments of culpability or collaboration in her book *Against the Unspeakable*, Hollaback!'s advocacy recodes the appearance of complicity in bystanding as a transformable relationship between perpetrator, victim and a community of witnesses. Hollaback! and other bystander-oriented movement initiatives see the complicity in bystanding as a space of potential agency. Here, complicity stands "in relation to this sphere on which judgment has yet to fall."[25] While complicity represents the condition of possibility for one's culpability and collaboration in an event, for Mandel and social-movement actors, complicity first signifies participation, a relation rather than a judgment.

To re-examine what constitutes the situation of bystanding and its technologies of seeing without rushing to judgment, we must see "complicity [as] a potentially productive site."[26] Without excusing complicity in violence, or trying to explain it away, I aim, alongside Mandel and movement organizations such as Hollaback!, to "disentangle the concept of complicity from the charge of collaboration and the verdict of culpability"[27] to understand how complicity situates one to act otherwise. If, following Mandel, we "redefine complicity as a series of actions on which judgment has yet to fall,"[28] we can see it as a social script of violence that can be broken, interrupted and re-written, as anti-rape activists have argued.[29] In this sense, complicity is "the condition of possibility for ethical engagement with the violence" that constitutes the subject position, and perspectival viewpoint, of the bystander,[30] a relational process that could end up somewhere else than collaboration in violence.

Hollaback! and other bystander-education initiatives deploy what Gilles Deleuze and Felix Guattari termed "the pedagogy of the concept," a way of redefining what being a bystander means and what bystanding constitutes as a form of participation in the context of their activism.[31] As the Hollaback! Winnipeg video blogs illustrate, bystander campaigns make teachable a set of targeted intervention strategies that require activists to see the bystander position differently. Rather than a problem, the bystander offers a solution to social violence and cultures of harassment. Hollaback! and other bystander-oriented advocacy groups develop a collective way of talking about gendered, racialized and sexualized forms of violence and harassment in which bystanding has become the focus of intervention strategies. In the process, they name new subject positions that come to replace that of the bystander: interveners, upstanders[32] and interrupters.[33]

Hollaback! and other campaigns such as "Not Your Baby," launched by the Metropolitan Action Committee on Violence against Women and Children in Toronto, ON, use mobile-phone applications and Google mapping technology to assist women, as well as queer, trans and racialized targets of harassment, in documenting street harassment as it is happening. "Not Your Baby" provides iPhone and android apps with ready-made verbal responses that people who witness sexual harassment can address to harassers to overcome the barrier that witnesses report of not knowing what to say in the moment of seeing, hearing and being the target of harassing encounters.[34] Both campaigns provide an activist vision of gendered media production that transforms what is normally conceived as the passive role of the bystander who witnesses harassment into an active documentarian of it. Hollaback! recently started another campaign, "I've Got Your Back," aimed at collectivizing responses to street harassment and offering online networks of support for those who report and document it, where people move from secondarily witnessing the testimony of others' harassment to turning their listening to that testimony into the capacity for collective support and action.[35] Hollaback! joined forces with the national Green Dot campaign, which uses the distributive viral metaphor of disease to re-code the activist imagination of disseminating bystander intervention strategies across the world.[36] Viewed through these campaigns, the bystander appears as a unique target for public opinion mobilization.[37]

Being a bystander signifies a capacity to act on behalf of others. Through Hollaback! one learns how to do so at a relatively safe distance from within the situation of witnessing. Users of Hollaback!'s mobile-phone application produce an account of the scene of harassment from the scene itself. They learn how to use the cellphone cameras surreptitiously and are advised to create some distance from the scene to record it safely. By database mapping their narratives and videos of photos through Google mapping technologies, others are then asked to bear witness to their account of the harassment. Elizabeth May describes the campaign:

> We knew that the only good way to provide real-time relief to people who are harassed is to get bystanders engaged, but we also knew that bystanders wouldn't act unless we showed them how. Our concept was this: we'd develop resources, trainings, and we'd start mapping bystander stories in green dots. Then, we'd build an "I've Got Your Back" button which users can click to show support. At the end of each day, the person who submitted their story will get an email telling them how many people have their backs.[38]

Through "I've Got Your Back!" others collectively stand by the targets of harassment, to offer at-a-distance support and participate in the mobilization of an anti-harassment movement emboldened with the networked tools of technological bystanders. In a study conducted by Cornell University, researchers found that people who posted stories of harassment on the Hollaback! site reported that "in cases where a bystander took action by confronting the harasser, the harassment was more likely to cease. Importantly, bystander interventions that had a positive influence on the target of harassment could be as simple as a knowing look or an empathetic statement that showed support."[39]

As their activism suggests, Hollaback! offers a key model of feminist activist video production that works to defuse harassment by turning the target of the gaze back onto harassers and mobilizing the support of bystanders. By database mapping their narratives and videos of photos through Google mapping technologies, others are then asked to bear witness to their account of the harassment, creating the possibility to collectivize responses to harassment, from the bystander as individual intervener to bystanders as a community. Movement organizations such as Hollaback! demonstrate that people have to learn how to intervene and they need tools that enable that learning. Intervention requires knowledge, training and habit building. Most people are ill-equipped to respond to emergency situations. Many of us do not learn to develop habits that would enable us to think quickly in emergencies.

As Elaine Scarry warns, while emergencies are met with calls to, "stop thinking and act," we need instead to develop habits of thought that provide us with "the power to anticipate and eliminate the conditions of emergency."[40] "Thinking," rather than being an impediment to action, "should bring about the end of emergencies."[41] Being able to intervene requires what Pierre Bourdieu called "practical knowledge," habituated ways of knowing that enable bystanders to act *as if* they are not thinking, "below the level of calculation and even consciousness, beneath discourse and representation."[42] In this light, the production of mobile-phone applications that enable bystanders to safely intervene in harassing situations performs a kind of activist-bystander pedagogy. The portability of cellphones and their improved photographic and video-making capabilities deploy bystanding as a productive, community-based and sometimes counter-state tool of documentation.

The crucial point of transition from bystander to interventionist witness lies in the move to document. While Hollaback! addresses the harms that bystander non-intervention can inflict on those who suffer harassment,[43] they approach bystanding as a position of possible intervention, a role that one can be trained to inhabit. The tools they provide by way of mobile-phone applications represent an extension of the ability to respond that is already structured into relations of bystanding. They enable mobile-phone users to intervene in a way that links the people targeted by harassment and those who see it happening in their positions as bystanders into a community of witnesses who stand behind them. They turn the cellphone from a tool of documentation into a technology of community-building.[44]

For Hollaback! the distinction between bystander and witness is less great than constructions of bystanding as a failure to bear witness suggest. According to performance studies scholar Caroline Wake, "[r]ather than establishing or reinforcing the distinction between active and passive spectatorship," as so many definitions of bystanding as failed or passive witness do, "theories of primary witnessing actually point [instead] to different modes or degrees of activity. . . . For the spectator positioned as bystander," Wake argues, "witnessing is both a conscious and self-conscious activity."[45] Through the lens of performance studies, how one plays the role of bystander is not fully determined or defined in advance. To be a bystander is to be in the position to be able to bear witness, to become response-able in relation to others, and in contexts that offer no guaranteed outcomes. Rather than pose that position as necessarily passive, then, Wake suggests that bystanding is participatory.

From the perspective of advocates against police brutality, bystanding often represents the only way one can safely respond to a scene of police violence: by documenting the scene as a bystander. A series of bystander cellphone videos from the early-morning hours of January 1, 2009 recorded a Bay Area Rapid Transit (BART) police officer, Johannes Mehserle, shooting transit rider Oscar Grant as he lay face down and handcuffed on the platform of the Fruitvale Oakland, California BART station. On YouTube, pages of cellphone videos taken by bystanders on the BART train at the time appeared in the days after the officer killed Grant, providing video evidence of the murderous shooting (Figure 2.4). On the audio tracks of videos, we hear witnesses exclaim in horror at the shooting and provide interpretations of the violence as police murder, suggesting that the role of bystander is that of both documentarian, interpreter and, by extension, witness. It was not safe for these cellphone-enabled witnesses to otherwise intervene in face of an armed police officer who had already killed a fellow rider. Their documentation was interventionist. The prosecution used mobile-phone videos recorded by other BART riders as evidence in their case against Mehserle, helping to convict the officer of involuntary manslaughter during his trial for murder.[46] In addition to being part of the record of police brutality and human-rights abuse, cellphone videos are sometimes the only existing documentation of such abuses, mobilizing social action and rebellion, in the case of the Grant shooting, which spurred protests in downtown Oakland.[47]

Mobile phones serve as key technological bystanders to popular movements and popular revolt.[48] In the context of popular revolutionary movements such as the 2010 protests in Iran, the 2011 Arab Spring movements and popular protest against government austerity measures witnessed in cities across the UK during August 2011, cellphones became essential tools of bystander intervention – recording violent security-force crackdowns on protesters and the shooting deaths of dissenters, on the one hand, and forms of social solidarity and helpful bystander behavior on the other. "Immersed in the crowd, yet able to communicate beyond it," cellphone users, as Vicente

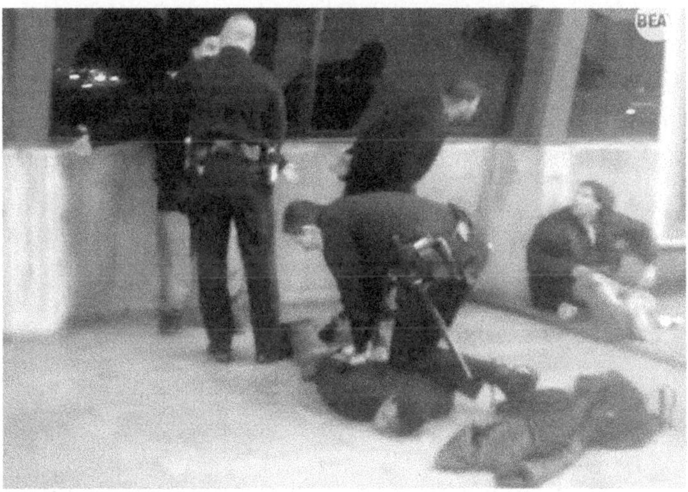

Figure 2.4 Still from a cellphone video of Oscar Grant's detention and shooting by BART police, January 1, 2010.

Source: https://youtu.be/Q2LDw5l_yMI.

Rafael argues, become constitutive of a larger body politic.[49] For archivists, cellphone recordings offer unique evidentiary and informational value about social change.[50]

In the present context of movement mobilization, the bystander position signifies a point of intervention and the potential location for social change agency. While the bystander typically represents someone proximate to an emergency who fails to intervene, constituting a *failed* (and passive) *subject*, in social movements, the bystander represents a potential intervener, an *active responsible subject.* My interest lies in the dis-articulation of (1) the bystander as a kind of person, a subject who fails because they do not exhibit feelings of care for others, from (2) bystanding as a position one occupies in proximity to events as they unfold, to (3) bystanding as a practice and embodied process of movement, hesitation, standing, looking and listening and, in some cases, intervening and offering help. This shift from the person of the bystander and the positionality of the bystander to the act of bystanding as a process of potential intervention focuses attention on bystanding as an act of witnessing that occurs in an unfolding situation.

The next part of the chapter examines a situational approach to bystanding through an analysis of bystander videos, both those recorded via closed-circuit television (CCTV) surveillance cameras and those documented on mobile phones. Video recording makes visible and audible the scene of bystander participation, defining bystander non-intervention as a social problem in ways that highlight the possibilities for intervention. Unlike photographs of bystanders, which pose the bystander as little more than an onlooker, video documentation of the bystander problem poses the bystander as someone who occupies a position in a scene of action.

Video evidence for the social problem of bystanding

Neither evidence of public apathy, social de-sensitization to violence or individual failures to feel or "feel badly," videos of bystanding represent a situational approach that poses the relationship between participants in the scene of bystanding as the key site of analysis and intervention. Within the moving images of digital video documentation (some with sound-recording capabilities and others without), the situation of bystanding unfolds in process through moving images rather than in the snapshot form of photography. Still photography has conventionally been used to depict bystanding as a practice of looking.[51]

Barbie Zelizer, Sharon Sliwinski and others have noted that photographs of bystanders to the liberation of the Nazi concentration camps and during the 9/11 terrorist attacks in New York powerfully define the act of bystanding through conventional forms of depiction and visible gestures and looks. As an act of citizenship training in how to see mass violence from the position of the bystander-witness, viewers of photographs of the Holocaust and of 9/11 are asked to judge images of bystanding subjects based on how those bystanders "look" and "look at" the scene they (often) inadvertently witness. The body of the bystander and his or her comportment and expression become physiognomic markers of their responsibility or irresponsibility for what they see. According to Zelizer's *Remembering to Forget: Holocaust Memory through the Camera's Eye*, after the liberation of the camps, photographs of witnesses to the atrocity focused on German nationals, soldiers and American political representatives as bystanders to the Nazi genocide against European Jews, Roma, homosexuals and the disabled, bringing onlookers into view through photographic documentation. The process of de-Nazification post-World War II involved the use of photography to encourage German nationals and international news audiences, particular among allied countries, to look at atrocity through photographs that included those of direct, non-survivor witnesses to the camps.

Post-9/11 photography also focused on witnesses to the burning World Trade Center towers and the people who jumped to their deaths. In viewing the massive archive of amateur photographs that constituted the exhibit "Here Is New York," Sharon Sliwinski describes a photographic convention for portraying onlookers in the "state of being transfixed": eyes wide open, hands over their mouths or wrapped around their heads.[52] Such conventionalized photographs of witnesses function as visual testimony for the experience of being witness, as onlookers to the collapsing World Trade Center towers and the people who jumped to their deaths from the collapsing, burning buildings. The "impact of the incident" for onlookers, she argues, "is expressed through bodily action and relayed through the medium of photography."[53] The body here becomes a point of transmission for feelings not otherwise articulated verbally, expressing the power to be moved via felt testimonies of the body and its gestures.

In media depictions and records of bystanding, one sees bystanders in at least two distinct ways: first, as subjects who look and who are depicted in the embodied process of looking; and second, as subjects who are part of a scene in which bystanding appears as a process, as a form of agency *in situ* through which judgment is formed. Bystander photography most readily conforms to the former way of seeing bystanders. By virtue of their moving-image formats and, for some, sound tracks, bystander videos draw attention to the situation in which bystanding occurs, emphasizing bystanding as a process more than a *kind* of "look." Bystander videos, I suggest, disarticulate seeing in proximity from understanding.

Fixed-location surveillance video and cellphone videography make bystanding audio-visible as a mediated "structure of relating, and not [yet] a charge (e.g. collaboration) or a verdict (e.g. culpability)."[54] Video documentation of bystanding functions less as a kind of visible evidence of the inaction of failed subjects and more as a mode of documentation that produces video technologies and their spectators as bystanders toward seemingly interventionist ends. In the context of emergency, bystanding represents a contingent and often heightened experience of crisis in process that tests individuals' and collectives' very ability to respond to emergency, and to know how to respond.

Bystander videos have become an active part of the social construction of the bystander problem and its redefinition as a site of social-movement intervention. Their aggregation and circulation within contexts of judgmental commentary online provides the context in which the work of transforming bystanding from "hearing and seeing in proximity" to bearing some responsibility to act within the situation on view is actively described and discussed. We might call these online sites of bystander video aggregation on YouTube a situational pedagogy of bystanding. Bystander videos that circulate online and in US television shows such as ABC's *What Would You Do?* screen bystanding around the potential for bystanders' social intervention, modeling forms of intervention in the process. They create spaces of judgment in which injustice and justice can be visualized and made audible.

Bystander videos are situational: they show scenes or situations of bystanding and they are effects of the situations in which they are recorded. In bystander videos recorded by CCTV and cellphone cameras, video cameras themselves become bystanders to the scene of bystanding they record. Via CCTV, scenes of bystanding are shot from above the bystander situation on cameras mounted on poles or the sides of buildings above the street. The camera in these bystander videos documents the scene of action and judgment. Rather than provide a participant view of bystanding, they instead provide a surveillance view structured by the infrastructures of video security. They also do not provide audio capabilities. CCTV bystander videos represent a politics of verticality – a view from above that purports to see the whole scene but does not offer any promise of intervention based on its depiction of the scene, like the high-rise view that sees the street, but from a distance far above its scene of action.[55] This view from above purports to offer an objective or disinterested view, a video capture of the real, not unlike the "god eye of science" Donna Haraway has critiqued.[56] Security cameras see the scene of bystanding from the perspective of policing and private security protection of property, a view interested in the propertied relation of owner and owned rather than the personhood of those it records.

Cellphone videos function less as surveillance technologies around the problem construction of bystanding and instead are often positioned as participant videos. Shot within or just proximate to the scene of bystanding, they produce video records of bystanding from a register horizontal to the scene. They offer voice-over capabilities, recording speech and verbal cues from the person recording the video as well as those witnesses whose voices are proximate enough to be picked up by mobile-phone microphones. Cellphone videos then provide an aural record of bystander witnessing in addition to visual testimony from the scene, an ability to see and hear activities from the street level.

CCTV recordings and cellphone videos produce different ways of seeing like and as a bystander. CCTV cameras offer fixed-position bystander views of scenes, e.g. properties to be secured, while cellphones function as mobile bystander technologies. The latter provide what Peter Maas[57] describes as the reality effect of shaky movements, low-resolution imaging, background noises and the aesthetic styling of amateur videography that has come to be identified with video witnessing. That is, they move and "hear," if you will, within the scene while CCTV cameras bystand above it, offering a position of judgment like a just-distant viewer. In this way, CCTV and cellphone cameras provide different positions from which to audio-visualize justice as a mediated practice, producing different registers of seeing, viewing and hearing like a bystander.

In recent years, online video aggregators such as YouTube have become key archives for video documentation of bystander non-intervention. Video evidence of bystander non-intervention proliferates across sites such as YouTube, where online videos make visible others' bystanding in the name of social judgment, if not legally binding justice. On the *New York Post* channel on YouTube, for instance, one can find CCTV footage of the April 2010 murder of Good Samaritan Hugo Tale-Yax, who lay on a Jamaica, Queens, New York sidewalk while dying from stab wounds he suffered in the process of intervening in a domestic dispute.[58] Several people pass him by; one apartment resident nearby came out on the sidewalk to take a cellphone photograph and then left. Recorded from a register just above the scene, the surveillance video record of the passers-by provides a bird's eye view that places the spectator at an elevated register, out of the space of possible intervention. The *New York Post* got access to the surveillance camera footage located just above the sidewalk where Tale-Yax collapsed, disseminating the video first via television broadcast and later on its YouTube channel "NYPost." The video asks of us "what would you do?" and many of us may desire to believe that we would act differently than those who walk by or the guy who snaps a digital photo.

But as former UN High Commissioner of Human Rights and Chief Prosecutor of the International War Crime Tribunal Louise Arbour warned in the mid-1990s, people want to see themselves as interveners, but when faced with a situation in which bystanders could intervene, they usually fail to do so.[59] She calls those of us who occupy this fantasy position "Armchair Schindlers," people invested in their own self-perception as interventionist cosmopolitan citizens who can act in ways others cannot and do not. The faith here is that one has a stronger will to help than others do. But as I am suggesting, will is only one piece of intervention. Intervention requires know how, of having models for how to intervene and in what ways depending on the circumstance. One is not born an intervener: one becomes one.

State and corporate power often work against this very will. Over 1.4 million viewers watched a single full-color YouTube video showing footage from a surveillance camera on October 13, 2011 in which a white van is recorded running over two-year-old girl Wang Yue two times on a busy market street in Foshan, China. Spectators are also positioned above and at an angle to the scene, out of the position of intervention, but able to see the full scale of the street scene as an elevated onlooker. Eighteen people are recorded walking, biking and driving by her, until a woman grabs her off the street and takes her to her parents working nearby.[60] As subsequent news coverage reported, many people did not stop to offer aid because doing so under the law would make one more liable for the death of this little girl. This suggests that there are state incentives not to take responsibility. In this way, the construction of the bystander problem can also shift responsibility off of the state and its representatives when it ought to be placed back on them.

Sometimes the bystander position is identified with prosecution, becoming a form of participation in criminal justice. On March 11, 2011, Brittany Norwood murdered her co-worker Jayna Murray at a Bethesda, Maryland Lululemon store next door to an Apple store, where a ceiling-mounted security video camera at Apple recorded two employees listening through the wall to victim Jayna Murray yell for help for over 30 minutes. A courtroom sketch based on the security-video recording of the two ear-witnesses listening up against the shared wall of the Lululemon and Apple stores and brought into evidence during the trial was circulated in news coverage of the case. Spectators see the scene from above, in a space of judgment not seemingly implicated in the action, or inaction, of the two subjects being recorded but instead identified with the legal prosecution of Jayna Murray.

Surveillance videos such as those of victims Hugo Tale-Yax and Wang Yue function as visual testimonials to bystander inaction. Collected and packaged online alongside commentary boxes and able to be replayed, they become forms of visible evidence that perform bystanding as a serialized and transitive process caught on tape and in digital file formats. Unlike still photographs of bystanders, which depict a static image of the bystander as onlooker where the act of bystanding is depicted as the eye-witness who looks at or away from a scene, bystander videos dramatize bystanding as an embodied process of movement, hesitation, standing, looking and listening and, in some cases, intervening and offering help. While their contextualization in online video aggregators alongside textual responses and video commentaries may appear to reveal the truth of bystanding as a clear-cut problem in which most bystanders appear not to act on behalf of others, those who are shown engaging in helping behavior depict and interpret bystanding as a process of incomplete, interrupted emergency thinking that structures bystanding and its experience.

In August 2011, as rioting broke out across cities in the UK in response to the police shooting of a young black man in Tottenham and drastic government austerity measures, a cellphone video taken of a wounded young man who had been knocked off his bike only to be robbed by a bystander went viral. The BBC and other news outlets re-broadcast the video in their riot coverage.[61] Cellphone video showed the injured backpack-wearing student of Malaysian origin, 20-year-old Mohd Asyraf Haziq, at first being helped by a racialized bystander, and then being robbed by a white-identified man while the first man starts to rifle through his bag, perhaps looking for identification (Figure 2.5). Video of this robbery became one of the most visible and talked about episodes in the days of rioting, racially coding the problem of bystanding around white male predation against a young racialized male student and his possible helper. While the moving images of the video seemingly show the transformation of a helpful racialized bystander into one who enables another man to commit robbery against the wounded boy (that is, by not stopping him), the female voice over codes the bystanders as perpetrators: "Are they actually helping him up? Oh my God. [pause] Are they going through his bag? Ahhhhh! He just took something from his bag. [pause] Dickhead. [inaudible]."[62]

The voice over performs the role of incredulous female judge of male bystander behavior. Incredulity may be the most audible response to bystanding in its problem construction, buttressed by the outpouring of comments on YouTube and other online forums that charge the helpful and possibly

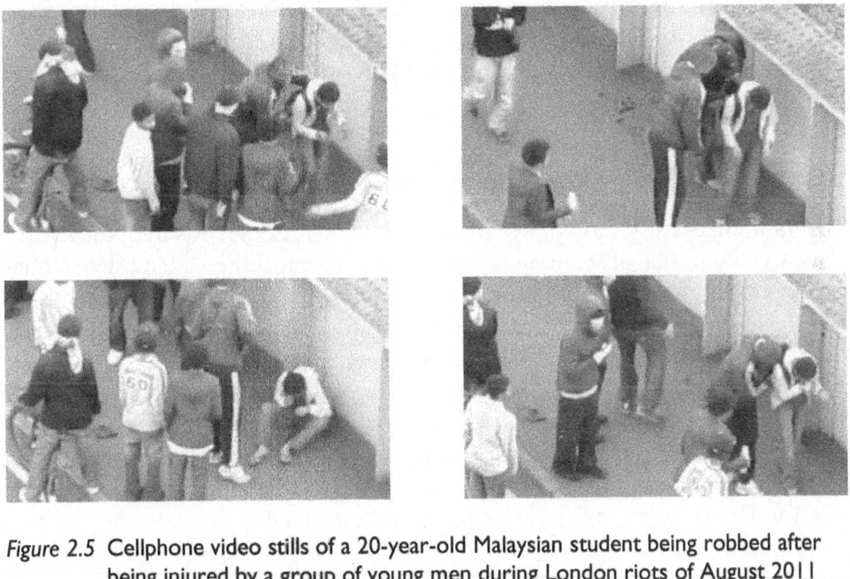

Figure 2.5 Cellphone video stills of a 20-year-old Malaysian student being robbed after being injured by a group of young men during London riots of August 2011 (YouTube video titled "Scumbag Rioters in London Fake Helping Injured Boy to Steal from Him").

Source: www.youtube.com/watch?v=oknjNw1elTo.

complicit bystander who is a young man of color with the same guilt as the somewhat older-appearing lighter-skinned man who steals what looks to be an iPad from the wounded young man's bag. In online commentary, the incredulous female judge's comments get coded in terms that cast each of the male bystanders as racialized folk-devils, in spite of their different behaviors dramatized on screen. The male bystanders in this video become "embodied allegories of inequality" that help legitimate the targeting of young men of color for heightened surveillance and policing.[63]

And yet the video shows a range of bystander behavior that complicates how bystanding is defined in either/or terms as victimization or perpetration. It instead appears as a contingent and changing position that shuttles between helping behavior, complicity and perpetration, illustrating Naomi Mandel's definition of complicity as, "a series of actions on which judgment has yet to fall."[64] Bystanding here signals a contingent relationship between responses to and participation in what one witnesses and responsibility for the actions one sees, a form of what Erving Goffman called "eye behavior on the street."[65] For Heather Love, this involves "the practice of re-description," where we reconceptualize "the social distribution of social attention on the street"[66] as a way of reading across the scene and the various positions of bystanding that unfold therein. While the online commentary surrounding the video demands us to judge the actions of each bystander similarly, as the actions of robbers, the video reveals a range of behaviors, some helping and some not, that diversify what viewers can see of bystander behavior in this particular moment. From a perspective on bystander intervention, it is worth holding onto a view that sees differences in bystander behavior.[67]

The "look" of taking responsibility

Bystanding is a situation in which video documentation and circulation forces one to take notice of seeming failures to act responsibly toward others. These videos and their infrastructures of dissemination constitute an emergent genre for representing the problem of bystander non-intervention in terms that make visible and audible the contingencies of occupying a bystander position and the call to take responsibility. The activities that shape and make up the bystander situation are not decided in advance, nor are the roles of participants fully determined, defined or scripted in generic terms. Lauren Berlant defines the situation as "a state of things in which *something* that will perhaps matter is unfolding amid the usual activity of life," a "state of animated and animating suspension."[68] The bystander situation represents what she terms a "genre of emerging event."[69] Later in *Cruel Optimism*, she describes the situation as a "disturbance, a *sense genre*"[70] of un-foreclosed experience in which time and action stretch out around the intensifying crises of ordinary life.[71]

Bystander videos call out for different conceptualizations of judgment and justice in the day-to-day. Justice, according to Sharon Sliwinski, requires spectators. Not only must justice be done, it must also be *seen* to be done: visibly performed for others to witness.[72] Today, TV shows such as ABC's *What Would You Do?* use hidden-camera observation techniques that combine the spectacle of "showing real people in unusual circumstances" with models of how not to be a bystander.[73] Through voice-overs, repeated playback of hidden-camera recordings and interviews between the show's host and unsuspecting bystanders, viewers of *What Would You Do?* are encouraged to judge other's non-interventionist behaviors and adopt more interventionist ones based on subjects who offered aid during the show's staged scenarios. Across its episodes, examples of injustice – defined as a failure to intervene in another person's harassment or discrimination – are illustrated through video replays that draw attention to the differences between when bystanders do and do not intervene on the show.

Shot using hidden-camera techniques made popular in earlier mid-twentieth-century TV programs such as *Candid Camera* and used in research experimentation and educational settings within the social sciences, dramatic hidden-camera bystander videos on *What Would You Do?* function as extra-legal, social evidence of the failures of justice and the possibilities for everyday just actions, as a process put under scrutiny by the audience. According to Anna McCarthy: "The value of hidden camera's social record lay in the way it [can] supplement liberal projects of reform and advocacy."[74] As revelatory media on the problem of bystanding, bystander video making on shows such as *What Would You Do?* dramatizes bystanding as a contingent process in which intervention into scenes of emergency, harassment, domination and violence *can* occur, even if they do not.

To move from bystanding as an experience of being proximate to an event that calls out for intervention to the act of bearing witness to it requires action in excess of seeing.[75] Sue Tait and Barbie Zelizer have asked what it means to bear responsibility for what people witness and to act in accountable ways. Does it require that one feels responsible or has the capacity to feel responsible? That is, is there an affect test that might determine what taking responsibility feels like, or ought to feel like? Or does taking responsibility refer to particular actions in which one's responsibility is audibly and visibly demonstrated? Like justice, does taking responsibility also need to be seen being done?

I ask these questions along with Tait and Zelizer not to dispute the need to take responsibility for others and to offer aid, but to open up a line of questioning that asks what it is to take responsibility as a demonstrable activity that, as bystander videos suggest, involves practices of documentation. Bystander videos seem to suggest that taking responsibility occurs in part through audio-visual reproduction of bystander situations, and the relations of judgment formed in audiencing them. The sense of witnessing I articulate here comes through the relationship audiences have to bystander videos. The video witnessing of bystander agency occurs post-fact, enabled by audio-visual recordings. Collected online, bystander videos seem to aggregate judgments against bystanders. Our very notions of taking responsibility for what one sees or hears are directly linked to the ability to record and distribute audio-visual evidence of bystanding and as bystanders.[76] While bystander videos provide opportunities to judge and hold bystanders accountable for their actions and inactions, audiences themselves become bystanders to the scenes and people on screen; they bystand the video bystanders.

From where, then, do our notions of what taking responsibility looks like come? If, as Sue Tait warns, the notion of bearing witness is too often conflated with ways of seeing and being an eye-witness, how do we come to know the extra work and learn the knowledge that makes up the practice of taking responsibility? What does taking responsibility look like? For many, taking responsibility appears most clearly in the verbal testimonial tradition, in the transition from witnessing to testifying: from seeing to saying. For Tait, "figuring a moral engagement with suffering requires a shift in emphasis from vision to voice expressed as response-ability."[77] John Peters[78] similarly described seeing as a passive act of witnessing while speaking signals action, marking out a difference between eye-witnessing and the testimonial tradition. Taking responsibility requires the "transmission of moral obligation" in a "discursive space that facilitates telling via empathetic listening."[79] As I have argued before, "people may simply not know how to act or what to do with their vicarious experience of others' suffering, because they have not been taught how to transform feeling into action."[80] To bear witness, I suggested, "*should* mean that citizens learn that mass acts of violence can continue to happen because so many bystanders have not been taught how to prevent violence."[81] Media coverage in itself cannot stop current and future atrocity, for "seeing does not necessarily compel responsibility."[82] In many cases, media witnessing "may actually not be about empowering citizens to act so much as it enables them to passively support state violence and the selective . . . commitment of humanitarian aid."[83] In this way, witnesses may actually be discouraged from if not actively prevented from acting otherwise.

Media-enabled witnessing, according to John Ellis, "involves certain elements of . . . that of the bystander"[84] that are currently being re-configured by social movements and educational initiatives. For Paul Frosh, electronic media "augment and transform what it means to witness"[85] through the organizational apparatus that produces and interprets audio-visual witnessing texts. These apparatus include social-movement organizations such as Hollaback! and others, such as Facing History and Ourselves. As Frosh argues, "Electronic media have multiplied the number of witnessed events reported to distant others, and multiplied greatly the number of those distant others. Most significantly . . . electronic media have altered the relationship between the witness and his or her addressees through the intervention of complex organizational and technical apparatus of audiovisual representation."[86]

In addition to bringing "citizens into direct encounter with images and sounds" of events,[87] witnessing poses a set of fundamental questions about communication in states of emergency, and the ability to respond in those situations. As Elaine Scarry[88] suggested after passengers on United Airlines flight 93 brought down their hijacked plane on September 11, 2001, mobile phones become key devices of collective citizen defense: bystander technologies put into the service of collective action. The increasing video documentary record of bystanding and the social-movement targeting of bystanders and bystander technologies requires us to reconsider the relationship between bystanding and media witnessing.

Bystander videos function as audio-visual witnesses to bystanders and to the situation of bystanding; they are both a record of the witnessed world and a means of reimagining witness that signals response-ability. Through this chapter, I have sought to untangle the association between bystanding, apathy and passivity through an analysis of how bystander videos depict bystanding as potentially interventionist ways of seeing, and how key social-movement organizations deploy mobile-phone video making as an activist-bystander practice. Bystander videos construct the problem by targeting onlookers as key participants in the events they witness or come in close proximity to – as subjects who proximately see events but often fail to act. The problem definition of bystanding, however, also carries with it another way of thinking about the bystander as someone who is positioned to intervene. While the construction of the bystander problem looks one way from the perspective of CCTV surveillance technologies, as a way of seeing like and with the state and corporate security, via Hollaback! the cellphone and its networks become a counter-tool of feminist surveillance and documentation, a way of seeing like a bystander in order to stand up to harassment. The cellphone can also be a tool of state-based seeing when Hollaback! reproduces racialized ways of seeing like the state by making men of color more visible as harassers than white men in their online documentation.[89]

As the bystander videos I have discussed here suggest, the construction of bystander non-intervention as a social problem and a target of intervention currently relies upon the technical capacity to record and display bystanding in action and on the scene via both fixed and mobile video. Current meanings of bystanding and bystander intervention are being defined via technological recordings of bystander behavior, through surveillance-camera footage shot from above the scenes of bystanding and cellphone videos shot from close proximity to the event of bystanding and often at the same scale of view as a nearby bystander. Both positions – that of the CCTV camera and the cellphone shot on the scene – are located within the range of proximity that Roger Silverstone[90] described as the moral and spatial dialectic of "proper distance." Proximity, for Silverstone, sets the basic conditions for ethical relations with others. While perfect connection with others, he suggests, is not possible, the ability we have to build moral relationships with others is grounded in the recognition of our failure to completely connect with those same others. "The acknowledgement of the inevitability of that failure,"[91] is a quintessential feature of the struggle to connect across routine failures of communication.[92]

Bystander videos reveal this aspect of ethical relationality in especially striking, humbling and sometimes politically despairing form.[93] Socially interpreted as examples of everything that is wrong with people's relationships with one another, particularly around stranger sociability, bystander videos also suggest that there might be some political utility to the bad feelings they deploy, for they have mobilized a pedagogy of response-ability among social-movement actors invested in defining bystanding as a capacity for intervention.[94] In mobilizing an activist-bystander perspective on violence there is still more work to be done to fully grapple with the possibilities for agency based in proximate social relations among strangers, the tools for recording that experience as testimonial practice, and the relations of powers that can be enacted and re-enacted without a strong anti-racist, anti-oppression framework-orienting practice. In training people to see like bystanders, groups such as Hollaback!, WITNESS and Facing History and Ourselves show us that the position of the bystander is one of potential agency in proximity.

Notes

1 See Karen Seidman, "What Happened to Woman in Viral Flood Video? She's Ok," *Montreal Gazette*, January 30, 2013, A6.
2 Supriya Dwivedi, "Voyeurism Is a Hallmark of Our Times," *Montreal Gazette*, January 31, 2013, A17.
3 See Kelly Oliver, *Witnessing: Beyond Recognition* (Minneapolis: University of Minnesota, 2001).
4 See also Sue Tait, "Bearing Witness, Journalism and Moral Responsibility," *Media, Culture & Society* 33, no. 8 (2011): 1220–1235.
5 Oliver, *Witnessing*, 15.
6 Paul Frosh and Amit Pinchevski, "Introduction: Why Media Witnessing? Why Now?," in *Media Witnessing: Testimony in the Age of Mass Communication*, ed. Amit Pinchevski and Paul Frosh (London: Palgrave MacMillan, 2009), 1–22.
7 Frosh and Pinchevski, "Introduction," 4.
8 See John Ellis, *Seeing Things: Television in the Age of Uncertainty* (London: I.B. Tauris, 2000).
9 Rick Prelinger, "The Appearance of Archives," in *The YouTube Reader*, ed. Pelle Snickars and Patrick Vondereau (Stockholm, Sweden: National Library of Sweden, 2009), 268–74.
10 Tait, "Bearing Witness, Journalism and Moral Responsibility," 1220–1235.
11 Ibid., 1227.

12 Peter Gabriel, Gillian Caldwell, Sara Federlein, Sam Gregory and Jenni Wolfson, "Moving Images: WITNESS and Human Rights Advocacy," *Innovations* 3, no. 2 (2008): 268–274.
13 See John D. Peters, "Witnessing," *Media, Culture & Society* 23: 707–723; see also Oliver, *Witnessing* and Tait, "Bearing Witness, Journalism and Moral Responsibility."
14 Peter Walker, "WITNESS: CCTV for the Masses?" *Innovations* 3, no. 2 (2008): 65.
15 Elizabeth Miller and Michele Smith, "Dissemination and Ownership of Knowledge," in *Handbook of Participatory Video*, ed. E.J. Milne, Claudia Mitchell and Naydene de Lange (Lanham, Md.: AltaMira Press, 2012), 331–348.
16 Paul Frosh and Amit Pinchevski, "Introduction," 11.
17 Emily May cited in Paul DeBenedetto, "MTA: If You See Groping, Say Something," NBCTV 4, New York. March 24, 2011, www.nbcnewyork.com/news/MTA-Institutes-New-Subway-Harassment-Policy-118587999.html .
18 Peters, "Witnessing," 709.
19 Ibid., 709.
20 Jodie Layne and Cleo Leslie, "An Introduction to the Bystander," video blog post (Winnipeg, Manitoba: Hollaback!, 2012), www.youtube.com/watch?v=HlrGxZCy9Hw.
21 Jodie Layne and Cleo Leslie, "Bystander Intervention," video blog post (Winnipeg, Manitoba: Hollaback!, 2012), www.youtube.com/watch?v=mOIcx5CHayI.
22 See Alison Young, *Imagining Crime: Textual Outlaws and Criminal Conversations* (Thousand Oaks, Calif.: Sage Publications, 1996).
23 Naomi Mandel, *Against the Unspeakable: Complicity, the Holocaust and Slavery in America* (Charlottesville: University of Virginia Press, 2006), 218.
24 Ibid.
25 Ibid, 214.
26 Ibid, 23.
27 Ibid, 29.
28 Ibid, 29.
29 See Sharon Marcus, "Fighting Bodies, Fighting Words: A Theory and Politics of Rape Prevention," in *Feminists Theorize the Political*, ed. Judith Butler and Joan W. Scott (New York: Routledge, 1992), 166–185. See also Martha McCaughey, *Real Knockouts: The Physical Feminism of Women's Self-Defense* (New York: New York University Press, 1997).
30 Mandel, *Against the Unspeakable*, 22.
31 Gilles Deleuze and Felix Guattari, *What Is Philosophy?* (New York: Columbia University Press, 1994), 16.
32 *Who Is an Upstander?* Digital video. (Brookline, Mass.: Facing History and Ourselves, 2011), www.facinghistory.org/video/heroes.
33 Steve James and Alex Kotlowitz, *The Interrupters*. DVD (Kartemquin Films, 2011).
34 See Desmond Cole, "New Mobile App 'Not Your Baby' Calls out Sexual Harassment," *The Torontoist* (September 13, 2012), http://torontoist.com/2012/09/new-mobile-app-not-your-baby-calls-out-sexual-harassment/; Sadie Whitelocks. "I'm Not Your Baby! The iPhone App that Gives Women the Perfect Withering One-liners to Answer Embarrassing Catcalls," *Daily Mail* (September 17, 2012),

www.dailymail.co.uk/femail/article-2204656/Im-baby-The-iPhone-app-gives-women-perfect-withering-liners-answer-embarrassing-catcalls.html.

35 Emily May, "I've Got Your Back Launches Today," press release, May 22, 2012, www.ihollaback.org/blog/2012/03/22/ive-got-your-back-campaign-launches-today/.

36 See Green Dot Etcetera, *Green Dot Map*, www.livethegreendot.com/map.html.

37 See William Gamson, "Bystanders, Public Opinion and the Media," in *The Blackwell Companion to Social Movements*, ed. David Snow, Sarah Soule and Hanspeter Kriesi. (Malden, Mass.: Blackwell, 2004).

38 May, "I've Got Your Back Launches Today.".

39 Beth Livingston, K.C. Wagner, Sarah Diaz and Angela Lu, *The Experience of Being Targets of Street Harassment in NYC: Preliminary Findings of a Qualitative Study of a Sample of 223 Voices Who Hollaback!*, report (Ithaca, N. Y.: Worker Institute at Cornell University, 2012).

40 Elaine Scarry, *Thinking in an Emergency* (New York: W.W. Norton, 2011), 10.

41 Ibid., 10.

42 Pierre Bourdieu and Loic Wacquant, "The Purpose of Reflexive Sociology (The Chicago Workshop)," in *An Invitation to Reflexive Sociology*, ed. Pierre Bourdieu and Loic Wacquant (Cambridge, Mass.: Harvard University Press, 1992), 128.

43 Livingston, Wagner, Diaz and Lu, *The Experience of Being Targets of Street Harassment in NYC*.

44 See also Michelle Caswell, "Instant Documentation: Cell-phone Generated Records in the Archive," *The American Archivist* 72 (2009): 138.

45 Caroline Wake, "The Accident and the Account: Towards a Taxonomy of Spectatorial Witness in Theatre and Performance Studies," *Performance Paradigm* 5, no. 1 (2009): 6.

46 See "Dramatic Video of BART Shooting Released by Court," blog, Los Angeles Times, June 24, 2010, http://latimesblogs.latimes.com/lanow/2010/06/dramatic-video-of-bart-shooting-released-by-court.html.

47 See Caswell, "Instant Documentation,"138.

48 See also Caswell, "Instant Documentation"; Negar Mottahedeh, "Iranian Women in Protest: 1953, 1978, 2009," *Scholar and Feminist Online* 10, no. 3 (2012), http://sfonline.barnard.edu/feminist-media-theory/iranian-women-in-protest-1953-1978-2009/; Setrag Manoukian, "Where Is this Place? Crowds, Audio-vision, and Poetry in Postelection Iran," *Public Culture* 22, no. 2 (2010): 237–263; Vicente Rafael, "The Cell Phone and the Crowd: Messianic Politics in the Contemporary Philippines," *Public Culture* 15, no. 3 (2003), 399–425; Elaine Scarry, "Citizenship in Emergency: Can Democracy Protect Us against Terrorism?" *Boston Review* 26, no. 5 (2002): 40–45.

49 Rafael, "The Cell Phone and the Crowd," 405.

50 Caswell, "Instant Documentation," 135.

51 See Barbie Zelizer, *Remembering to Forget: Holocaust Memory through the Camera's Eye* (Chicago: University of Chicago Press, 1998); Sharon Sliwinski, *Human Rights in Camera*. (Chicago: University of Chicago Press, 2011).

52 Sharon Sliwinski, "New York Transfixed: Notes on the Expression of Fear," *The Review of Education, Pedagogy and Cultural Studies* 30 (2008): 236.

53 Sliwinski, "New York Transfixed," 237.

54 Mandel, *Against the Unspeakable*, 216.

55 See Jane M. Jacobs, Stephen Cairns and Ignaz Strebel, "Materializing Vision: Performing a High-Rise View," in *Visuality/Materiality: Images, Objects and Practices*, ed. Gillian Rose and Divya P. Tolia-Kelly (Farnham, Surrey: Ashgate, 2012), 133–154.
56 Donna Haraway, "Situated Knowledges: The Science Question in Feminism and the Privilege of Partial Perspective," *Feminist Studies* 14, no. 3 (1988): 575–599.
57 Peter Maas, "Perpetrators and Participants: War Photographers of the Digital Age," Talk at the international symposium *Conflicted Reporting: Journalism Then and Now*, Montreal: McGill University, November 2, 2012.
58 "Hero, Homeless Man Stabbed to Death – New York Post." Digital video. New York Post YouTube channel, April 24, 2010, www.youtube.com/watch?v=j-20dURI9qA.
59 Louise Arbour, "Progress and Challenges in International Criminal Justice," *Fordham International Law Journal* 21, no. 2: 531–540.
60 See "WTF?!?! 2YR Chinese Girl Ruthlessly Run Over Twice and Pedestrians Do Nothing" (2011). YouTube video, hwww.youtube.com/verify_age?next_url=/watch%3Fv%3DUqVYUzHc5L8.
61 See "UK London Riots 2011 – Victim 'Feels Sorry' for Looters who Robbed Him," BBC news video, United Kingdom News YouTube channel, www.youtube.com/watch?NR=1&feature=endscreen&v=44BCvUsswFQ; "London Riots – Scum/Looters Steal from Injured Boy, Recovery," BBC News video, United Kingdom News YouTube channel, www.youtube.com/watch?v=J0luHzT30d8.
62 The person who uploaded the video on August 9, 2011 credits the video's recording and original upload on Facebook to Abdul Hamid; other YouTube uploads also credit Hamid. The incident occurred in Barking, London.
63 Jacques Rancière, *The Emancipated Spectator* (New York: Verso, 2009), 12.
64 Mandel, *Against the Unspeakable*.
65 Erving Goffman, *Relations in Public: Microstudies of the Public Order* (New York: Harper Colophon Books, 1971), 24–26, quoted in Heather Love, "Close but Not Deep: Literary Ethics and the Descriptive Turn," *New Literary History* 41 (2010): 380.
66 Love, "Close but not Deep," 379, 381.
67 Thanks to Nic Sammond for an especially fruitful conversation about seeing the differences in bystander behavior within this video back in August 2011.
68 Lauren Berlant, *Cruel Optimism* (Durham, N.C.: Duke University Press, 2011), 5.
69 Ibid.
70 Ibid, 5.
71 Emphasis added; see ibid., 5, 195.
72 Sliwinski, *Human Rights in Camera*, 4.
73 Anna McCarthy, "'Stanley Milgram, Allen Funt, and Me': Postwar Social Science and the First Wave of Reality TV," in *Reality TV: Remaking Television Culture*, ed. Susan Murray and Laurie Ouelette (New York: New York University Press, 2004), 19–39.
74 Anna McCarthy, "'Stanley Milgram, Allen Funt, and Me'," 21.
75 See Zelizer, *Remembering to Forget*; Tait, "Bearing Witness, Journalism and Moral Responsibility."
76 Thanks to Alison Fyfe, PhD student in Communication Studies at McGill, for pointing out this feature of bystander witnessing via audio-visual reproduction.

77 Tait, "Bearing Witness, Journalism and Moral Responsibility," 1222.
78 Peters, "Witnessing," 709.
79 Tait, "Bearing Witness, Journalism and Moral Responsibility," 1227.
80 Carrie Rentschler, "Witnessing: U.S. Citizenship and the Vicarious Experience of Suffering," *Media, Culture & Society* 26, no. 2 (2004): 300.
81 Rentschler, "Witnessing," 302.
82 Tait, "Bearing Witness, Journalism and Moral Responsibility," 1226.
83 Rentschler, "Witnessing," 302.
84 John Ellis, "What Are We Expected to Feel? Witness, Textuality and the Audiovisual," Screen 50, no. 1 (2009): 75.
85 Paul Frosh, "Telling Presences: Witnessing, Mass Media and the Imagined Lives of Strangers," in *Media Witnessing*, 50.
86 Paul Frosh, "Telling Presences," 50.
87 Ellis, Seeing Things: Television in the Age of Uncertainty, 9.
88 Scarry, "Citizenship in Emergency."
89 Jessie Daniels, "Rethinking Cyberfeminism: Race, Gender, and Embodiment," *WSQ: Women's Studies Quarterly* 37, nos. 1&2 (2009): 101–124. In October 2014, women of color such as Roxane Gay (via Twitter) and Colorlines' Akiba Solomon called out Hollaback! for the video they posted on YouTube, "10 Hours of Walking in NYC as a Woman," which films a white-identified woman walking through Harlem and being harassed by several men of color. Rob Bliss of the PR firm that made the film for Hollaback!, Rob Bliss Creative, explained the decision to edit out the white men they recorded by saying the sound was inaudible and the visuals were obstructed. See Akiba Solomon, "On that Street Harassment Video and Race," *Colorlines: News for Action* (October 30, 2014), http://colorlines.com/archives/2014/10/on_that_street_harassment_video_and_race.html.
90 Roger Silverstone, "Proper Distance: Towards an Ethics for Cyberspace," in *Digital Media Revisited: Theoretical and Conceptual Innovations in Digital Domains*, ed. Gunnar Liestol, Andrew Morrison and Terje Rasmussen (Cambridge, Mass.: MIT Press, 2004), 469–490.
91 Roger Silverstone, "Proper Distance," 483.
92 See also John D. Peters, *Speaking into the Air: A History of the Idea of Communication* (Chicago: University of Chicago Press, 1999).
93 See Wendy Brown, "Resisting Left Melancholy," *Boundary* 2 26, no. 3 (1999): 19–27; Heather Love, *Feeling Backward: Loss and the Politics of Queer History* (Cambridge, Mass.: Harvard University Press, 2007); and Lisa Duggan and Jose Munoz, "Hope and Hopelessness: a Dialogue," Women & Performance: A Journal of Feminist Theory 19, no. 2 (2009): 275–283.
94 Todd Carmody and Heather Love, "Try Anything," *Criticism* 50, no. 1 (2008): 133–146; Heather Love, "Feeling Bad in 1963," in *Political Emotions: New Agendas in Communication*, ed. Janet Staiger, Ann Cvetkovich, and Ann Reynolds (New York: Routledge, 2012): 112–133.

Chapter 3

Professionalizing police media work

Surveillance video evidence and the forensic sensibility

Kelly Gates

The last several decades saw widespread diffusion of closed-circuit television (CCTV) systems in urban areas, workplaces, retail outlets, banks, casinos, roadways, airports, and other public and private spaces. This diffusion was most prevalent in countries experiencing ascendant neoliberal politics during this period: the dismantling of social welfare programs; deregulation of industries; privatization of government programs; and a redirection of economic and social policy toward corporate-capitalist priorities.[1] The vastly increased use of CCTV for monitoring public and private spaces has introduced a host of problems for the management of these systems and their enormous output of visual media. This proliferation of CCTV systems and the video content they generate also raises important questions for the critical-cultural analysis of policing and the way visual media technologies shape, and are shaped by, police work.

This chapter looks at police efforts to optimize CCTV functionality, not for real-time monitoring but for the post-crime analysis and evidentiary uses of surveillance video. In the emerging field of *forensic video analysis* (FVA), we find a set of technologies, practices, and professional standards taking shape to exploit the evidentiary potential of a vastly increasing volume of recorded surveillance video. While the use of photographic media for evidentiary purposes is nothing new to police practice, the proliferation of video surveillance has generated unique issues in the domain of visual evidence. Exploiting the evidentiary potential of surveillance video requires novel types of *new-media work*, and a critical question concerns the relationship between this new-media work and the strategies the police employ to gain and maintain narrative authority and legitimacy. How are the symbolic and material dimensions of policing and police authority being transformed in the new-media landscape? On the one hand, new media provide ways of challenging the narrative authority that the police have conventionally held. A number of observers have noted that new handheld devices, combined with distributed digital networks and social media platforms, are investing ordinary people with new ways of visually documenting police conduct and circulating the media they produce.[2] For example, Andrew Goldsmith has examined the implications of new-media practices for "police image management and accountability," arguing that the new "uncontrolled visibility" of the police poses challenges for their ability to control their image and maintain authority.[3] On the other hand, my focus here is on the ways the police use new-media technologies to support their own narrative strategies, giving them new tools to do the "knowledge work" that they have long been charged with performing.[4]

Elsewhere I have argued that the status of video evidence as an index of real events – a sign or representation that offers a direct, empirical connection to material reality – is the result of an intentional process of production.[5] This process involves the repurposing of new technologies borrowed from the domain of creative media production in order to transform a chaotic field of raw surveillance video into useable evidence. In addition to the exchange in technologies, an unavoidable epistemological and interpretive exchange takes place between evidentiary uses of surveillance video on the one hand, and the now prevalent forms of surveillant narration found in both fictional and true crime storytelling on the other.[6] But despite this exchange in meanings and technologies, or partly because of it, considerable effort has gone into establishing formal standards for the evidentiary uses of surveillance video that distinguish the discovery of video evidence from the production of creative content. The status of FVA as an objective science requires differentiating it from the domain of creative video imaging and nonlinear editing techniques, even as these domains rely on overlapping technologies and ways of seeing and interpreting visual media.

Here I address another dimension of the effort to establish the scientific and legal credibility of FVA and the production of video evidence. I argue that the legitimacy of this field depends fundamentally on the *professionalization* of its practitioners. Professionalization aims to invest those who work with recorded surveillance video with a "forensic sensibility": not just a set of technical skills but also a way of seeing, thinking, and speaking professionally that articulates the practices and products of video analysis in terms of forensic science and the legal norms governing the use of photographic and motion-picture evidence. This forensic sensibility not only draws on a long history of investigative norms and procedures, but also includes a somewhat more recent focus on instilling in the forensic specialist a form of *computational thinking*. The emerging field of FVA is one site where an epistemic virtue of "computational objectivity" is taking shape: the belief that neutral scientific image analysis can be achieved by translating certain forms of professional trained judgment into computational processes.[7] But as I explain in this chapter, computational objectivity does not entail the *complete* automation of all forms of trained judgment. In practice, computational objectivity also involves investing trained professionals with computational thinking, or *computational vision*. Becoming a forensic video analyst entails learning how to look at images with a computational eye. It is by acquiring computational vision, applying and communicating that interpretive perspective, that forensic video analysts establish their own professional credibility and the status of their field as a legitimate science. In this way, an emerging professional community of law-enforcement video specialists aims to establish that FVA, when applied correctly and with professional integrity, reveals the truth about objects, people, and events depicted in recorded surveillance video.

Legitimizing FVA as a field of professional expertise

The spread of CCTV has meant that recorded surveillance video has become one of the most prolific sources of evidence in criminal investigations. One of the first things the police do now when arriving at the scene of a crime is to look for cameras.[8] According to the director of research for the International Association for Police Chiefs, "It used to be you got to a crime scene and what you had was whatever was left there: a cigarette stub or a tire skid . . . Now it's possible to have between 5 and 10 video clips that [police] can gather from that area."[9] Jonathan Hak, a crown prosecutor from Alberta, Canada, noted some time ago that it is "routine now for police to seize video from places within a mile of where a crime occurred."[10] Although difficult to measure precisely, a growing volume of criminal investigations includes video evidence. And while the evidentiary value of recorded surveillance video would seem self-evident, in fact the haphazard diffusion of CCTV – and the widely varying forms these systems take – has introduced major challenges for its effective use as an investigative and evidentiary technology.

In his history of photographic evidence, John Tagg notes the growth of specialized police photographers in England in the early twentieth century, following "the successful development of Sir Edward Henry's system of identification by means of fingerprints."[11] (Photography proved the best way to record finger impressions found at crime scenes.) Before this, the work of police photography was most likely "carried out by professional photographers who were not yet members of the police force itself."[12] Here, Tagg points to a domain of professional specialization in policing, one tied to developments in photography.[13] In fact, police historians have often connected the professionalization of policing to the adoption of technological innovations – photography as well as things such as fingerprinting, filing systems, the call box, patrol wagons, and standardized pistols.[14] Each innovation carries with it the requirement that the police develop and acquire new sets of skills and practices, in turn altering the range of responsibilities and activities that constitute the police work. Still, as Peter Manning argues in a recent study of new crime-mapping techniques, certain aspects of police work have remained remarkable consistent over time, even in the face of the police's adoption of new technologies.[15]

Today, with the growing use of video evidence in criminal investigations, police use of visual media is going through another transition of the sort described by Tagg, in complex relationship with a new wave of professional specialization. New kinds of visual-imaging expertise are taking shape partly within, and partly outside, the official ranks of the police. These forms of expertise aim to give the police more real-time visual access to their target areas for purposes of controlling those spaces, as well as more evidentiary ammunition and narrative authority in both the legal system and in public discourse about crime and policing.[16] Whether those performing the media work associated with video evidence are technically police officers or not (sometimes this work is outsourced to private contractors), the authority of the emerging domain of forensic video analysis – and especially its status as a legitimate investigative practice that produces "objective" results – depends fundamentally on the professionalization of its practitioners. Professional specialization in the area of FVA is related to, and an incidence of, a broader set of transformations in police work as a set of activities, an occupation, and a domain of knowledge production – as a social institution with a particular relationship to *truth*.

Without a doubt, FVA requires a considerable amount of technical expertise and knowledge of video hardware and software, far beyond pressing play, pause, rewind, and fast-forward on a video player. The Law Enforcement and Emergency Services Video Association (LEVA) defines FVA as the "the scientific examination, comparison, and/or evaluation of video in legal matters."[17] LEVA's official definition provides a starting point for explaining the field; however, exploiting the evidentiary potential of recorded surveillance video is not limited to the "examination, comparison, and/or evaluation" of video images. It also includes the *field acquisition* of video, as well as its *archival management*. Both field acquisition and archival management of video (also known as "media-asset management") are complex activities, requiring specialized skills. Individuals tasked with various stages of video acquisition, analysis, and media-asset management require an enormous and ever-changing knowledge base and set of technical skills to perform these activities.

LEVA's carefully crafted definition of FVA does more than provide a basic description of the field; embedded in this definition is an effort to make an authoritative statement about the field's scientific and legal standing. Individuals enrolled in LEVA's Level 1 training course, "Forensic Video Analysis and the Law," are required to memorize the definition as part of their introductory training, incorporating it into both their professional identity and the public presentation of that identity (especially when asked for a description of their expertise in court).[18] In LEVA's definition, and in the specialized forms of expertise being formulated for the production of video evidence, we find considerable emphasis on the scientific status of the field, and on the need for, and value of, objectivity.

This preoccupation with establishing the FVA's scientific objectivity is not unique to video forensics. It is widely recognized, in the legal system itself and other public forums, that forensic scientists in general face considerable pressures to produce results supportive of the demands of the investigative process. In other words, the forensic sciences are considered to be plagued by a greater potential for biasing influences than the research sciences. These issues were the focus a much-debated report from the National Academy of Sciences in 2009, "Strengthening Forensic Science in the United States: A Path Forward." The report identified systemic problems from the local to the federal levels (in the worst cases contributing to wrongful convictions), and called for a more rigorous culture of science in the forensic fields.[19] The work of science studies, a significant body of literature, much of it focused on issues associated with DNA evidence, has examined forensic science as distinct from research science, especially in terms of its relationship to questions of expertise, evidence, and truth.[20] Among this work, Simon Cole has argued that the "epistemic culture" of forensic science is very different from that of research science, and expecting the former to mirror the latter is an unrealistic goal.[21] More so than other domains of science, the forensic sciences have an often-tenuous claim to scientific credibility.

The US Supreme Court has also addressed the struggle for scientific credibility on the part of the forensic sciences.[22] In *Melendez-Diaz v. Massachusetts* (2009), the Court was asked to determine whether the results of forensic analysis could be admitted as evidence without testimony from an expert witness. The prosecutor in this case argued that the certificates of state laboratory analysts were the result of a technical process that produced factual information that should speak for itself, without the testimony of a qualified expert. The Supreme Court disagreed. In its ruling, the majority held that certified reports from forensics analysts were inadmissible as evidence unless the opposing side would have the chance to cross-examine authors of those reports in court. (The decision not only upheld the Sixth Amendment confrontation clause, but also relied on what may in fact be a misplaced faith in the adversarial trial system to serve as a means exposing flaws in forensic science.) In explaining its decision, the Court noted that most forensic laboratories are administered by law-enforcement agencies, introducing a special potential for bias. At the very least, the Court explained, "[b]ecause forensic scientists often are driven in their work by a need to answer a particular question related to the issues of a particular case, they sometimes face pressure to sacrifice appropriate methodology for the sake of expediency."[23] In the worst-case scenario, the Court noted, "a forensic analyst responding to a request from a law-enforcement official may feel pressure – or have an incentive – to alter the evidence in a manner favorable to the prosecution."[24]

Writing for a publication called *Government Video*, one observer explained what the *Melendez-Diaz* decision meant for video forensics, namely, that the field needs more rigorous and intensive professionalization.[25] Specifically, agencies need to dedicate more resources for ongoing technical training as well as training in courtroom performance.[26] The writer, himself a forensic video analyst, conceded that video evidence sometimes makes it into court to support claims that "cannot be supported by objective scientific processing methods."[27] "It isn't that video techs lie," he explained, "they just do not have the technical training or experience, in many instances, to see past what the ordering agency wants them to see."[28] Those who process video evidence need both a greater depth of knowledge of how their computational tools work, as well as an ability to explain the techniques *clearly* to non-experts, he insisted. Since "attorneys facing strong video evidence may push this ruling to extremes to get the video excluded," argued the author, "every forensic video technician in the field, the lab or private practice needs to immerse themselves in the technology of video and their tools."[29]

According to these comments from the field, the legitimacy of FVA requires qualified experts – trained individuals with a set of technical skills that enables them to produce useful and admissible video evidence. And becoming a professional forensic video analyst means developing the capacity to resist the persistent internal pressures that can introduce bias into one's work. It also means learning how to *present* the process of FVA, and the results, in a way that helps establish their credibility.[30]

Professional association, training, and certification in FVA

FVA is not a fixed and unified set of methods, skills, and technologies that individuals simply acquire and adopt in order to become qualified experts. Instead, it is an evolving set of practices and standards that involves the constant input of many actors – expert and novice, human and non-human – within and outside the official ranks of the police. The full range of skills and activities that constitutes FVA as a domain of expertise is not fixed but in the process of negotiation. FVA is characterized by constant change: changes in technologies and practices, as well as changes in the range of ethical considerations, legal strategies (prosecutorial and defensive), and investigative and evidentiary issues that come into play. Like other fields of scientific expertise, part of being an expert analyst involves keeping abreast of new developments. For the sake of discussion, it makes sense to distinguish between the professionalization of forensic video *analysis* on the one hand (its formation as a professional field of forensic science), and the professionalization of forensic video *analysts* on the other (the process by which individuals become trained experts). But this distinction is a tenuous one, with these two processes continuously feeding back into each other.

One source and advocate for both forms of professionalization is an organization already mentioned: LEVA, formed in the 1990s. LEVA's official mission is "to serve as a key resource to the global public safety community by focusing on the needs of video production and forensic imaging disciplines by providing opportunities for professional development through quality training and informational exchange."[31] LEVA began offering training in FVA in 2000. While most of this training is based in the United States, the organization has international representation. One of LEVA's primary self-defined responsibilities is to provide professional certification in the work of FVA. Official certification not only invests individuals with authority as experts in the field, but also the process itself involves adopting a professional self-identity and participating in the formation of a professional community. A description of the certification program states explicitly the relationship between professionalization and the scientific and legal legitimacy of FVA:

> As with all "new" sciences, the courts must be satisfied that the science is technically sound and that the witness using the science is properly qualified. . . . *LEVA's Certification Program assists in further establishing the legitimacy of forensic video analysis* and will help to ensure a positive reception of the analyst's evidence in court. . . . Certification is an important objective standard by which to judge the competence of analysts.[32]

LEVA's informational materials and activities provide a wealth of insights about the professionalization of FVA. However, individuals may perform the work of video forensics without LEVA training, and it is not necessary to have an official LEVA certificate to qualify as an expert witness in court. In the United States and Canada, one's authorization to provide expert testimony in a legal setting is typically determined in each particular case by the presiding judge. (The ability of judges and other triers of fact to adequately assess scientific expertise and expert testimony is a subject of extensive debate.) Nevertheless, an emerging community of experts aims to establish and control the boundaries of expertise in this domain, and one avenue for doing so is by strategically positioning LEVA as the go-to professional organization for FVA expertise, and by establishing LEVA certification as the gold-standard professional credential. Some LEVA members view the wide variability for what might count as expertise as a potential threat to the legitimacy of the field. (The term "wedding videographers" is sometimes used, derogatively, in reference to those considered "quacks" – people viewed as having no real expertise but who nonetheless do video work for the police and even qualify as expert witnesses in court.) The formation of LEVA, the recruitment of members, the development of training courses, certification requirements, regular conferences, a generative website for sharing information, and other activities that constitute the organization and its self-defined responsibilities and boundaries, are central to the formation of FVA as a domain of legal and scientific expertise.

Computational objectivity

The scientific and legal standing of FVA involves the formation of a professional community with an identity and a set of expert activities and qualifications. It also requires the field and its members to address the kinds of epistemic challenges associated with the scientific analysis of images, historically and in the contemporary transition to digital imaging. Each field of forensic science, from handwriting and bite-mark analysis to DNA identification, faces unique challenges in making credible claims to objectivity. For its part, FVA is subject to special scrutiny given common assumptions about the new level of image manipulability afforded by digital imaging techniques. The potential for images to be altered is generally thought to be much greater in the so-called post-photographic era, requiring those who make evidentiary use of images to establish the integrity of the techniques they apply to images, as well as the authority of the truth-claims they make with and about those images.

One way of addressing the challenge of digital image manipulability, as we will see, involves resorting to digital techniques themselves. While digital techniques are commonly viewed as a threat to image credibility, actors invested in preserving the principle of photographic truth have turned to algorithms and computation in an effort to bolster claims to objectivity associated with visual evidence. This turn to computational techniques as an avenue to achieving the "objective" analysis of digital images represents a recent move in a much longer history of scientific objectivity.

In their historical study of the prevailing "epistemic virtues" that have defined objectivity over time, Lorraine Daston and Peter Galison use the term "mechanical objectivity" to refer to the type of objectivity associated with photography and other visualizing instruments developed in the nineteenth century.[33] Proponents of "mechanical objectivity" subscribed to the belief that mechanical devices could be used to produce the scientific images that were uncontaminated by interpretation, in contrast to the artistically rendered, "true-to-nature" illustrations that populated scientific atlases. Photography promised to remove the individual scientist's judgment, and the biasing hand of the illustrator, from scientific image making. But photography and other mechanical visualization techniques never made good on this promise – the problem of image interpretation persisted – and what emerged in the twentieth century, as an acceptable avenue to objectivity, was the epistemic virtue of "trained judgment." In this view, objectivity could best be achieved through the analytical abilities and interpretive skills of well-trained human beings.

In the case of FVA, what we see emerging is another kind of objectivity, what we might call computational objectivity. This view of objectivity holds that the way to achieve an objective analysis of images is through computational processes, including both translation of forms of human perception into algorithms or computational systems, and the development and adoption of away of looking at and analyzing images by adopting a computationalist perspective. What I am calling computational objectivity involves the development of algorithmic techniques used to assist and enhance image analysis, as well as the perceptual labor of trained human beings – professionals with specialized skills that allow them to examine digital images with a thorough technical understanding of how the images were generated and what computational techniques are available to enhance their evidentiary value. Like the early form of "mechanical objectivity," computational objectivity does not and cannot involve the complete automation of analytical techniques. The role of human perception and professional trained judgment not only persists, but also is itself being pushed in new directions, with new types of expertise taking shape in complex relationship with evolving professional norms, legal standards, technologies, and forms of labor.

Digital image authentication and computational objectivity

One area of research and development associated with FVA that sheds light on what I am calling computational objectivity is *digital image authentication*, a set of visual and computational techniques being developing with the aim of determining whether an image has been altered in some way. Hany Farid, Professor of Computational Science at Dartmouth, New Hampshire, is considered a pioneer in this area. Farid's approach to creating tamper-detection software tools "starts with understanding what statistical or geometric properties of an image are disturbed by a particular kind of tampering."[34] He and his colleagues then "develop a mathematical algorithm to uncover those irregularities," coming up with algorithms that can identify things such as inconsistent lighting, "cloned" areas in an image, and composite images.[35]

As experts in this area emphasize, it is only possible to establish positive proof of image tampering; it is not possible to provide definitive proof of an image's authenticity. This is because there is always a possibility that an image has been tampered with, but the alterations are so sophisticated as to be undetectable using the available techniques. Like the field of FVA, new techniques and software programs for assessing image authenticity are themselves *in development* – in other words, they are in the process of being designed, published, and peer-reviewed by researchers and practitioners in this field. Digital image authentication is typically applied to still images, but the techniques fall within the purview of video forensics because analytical techniques typically involve examining video for visual traces frame-by-frame, as a series of still images. Video images are often separated into individual frames for analysis and evidence production, and for presentation in court.

Digital image authentication was the focus of a workshop at a LEVA training conference held in San Diego, California in October 2012.[36] The workshop framed the problem of image authentication within a broader discussion of professional ethics, beginning with a discussion of the biasing pressures that come into play in the work of forensic image analysis. Attendees – police officers and forensic specialists with varying levels of technical expertise, mostly from local law-enforcement agencies in the United States – were given ethical instruction on the virtues and need for objectivity, before any discussion of image tampering or authentication. Speakers – including faculty members from the National Center for Media Forensics at the University of Colorado – highlighted strategies for mitigating bias in forensic image analysis, including the need to adopt standardized methodologies where possible, as well as an objective "state of mind" and a sense of ethical responsibility. Significantly, one of the strategies the workshop emphasized as a means of mitigating bias – not only in digital image authentication but also in forensic work more broadly – was to employ computational forms of image analysis.

The available computational techniques for digital image authentication fall into four categories. First, file structure analysis involves investigating the different types of metadata associated with an image, for example, looking at the file structure to determine whether there are traces or signatures left by photo-editing software – a sure sign that the image has been changed in some way, although perhaps not enough to render it "inauthentic" or inadmissible as evidence. A second category involves global analysis, or examining the characteristics of an image as a whole, in contrast to a third category, local image analysis, or techniques that focus on specific, localized places on an image. Finally, source-specific forms of analysis involve identifying or verifying the imaging device that created the image, for example, by looking for pixel defects and connecting those to a specific imaging device. A forensic analyst might perform a combination of techniques to authenticate an image, and then interpret the results and offer conclusions based on whether the data suggest that the image is authentic, whether the results are inconclusive, or whether there is evidence of image tampering.

Significantly, while computational techniques are central to image authentication, the visual inspection of images by analysts, using their own perceptual capacities, remains central to the process of determining whether images have been altered. Analysts must be able to visually examine images with their own eyes to identify signs of tampering, knowing as much as possible about the available techniques of digital image manipulation. In other words, being an expert analyst capable of detecting image tampering means being able to look at and analyze images with a computational eye.

Jennifer Mnookin's historical analysis of the fraught introduction of photographic evidence in the US court system in the nineteenth century is instructive for understanding the importance of new digital image authentication techniques to the emerging epistemic virtue of computational objectivity.[37] As Mnookin observes, the problem of photographic manipulability was immediately apparent to the courts, and photography's privileged status as evidence was by no means self-evident. Mnookin focuses on a high-profile case in which photography's evidentiary status was hotly contested: in the fraud trial of the avowed "spirit photographer" William Mumler. In Mumler's portrait photographs, living persons appeared with visible apparitions of their dear departed hovering behind them. For some contemporary observers, Mumler's images provided convincing evidence for the claims of the spiritualist movement. But for the more rationally minded, the ghost photos were obvious hoaxes that posed a grave threat to the empirical authority of the new medium. Among the latter were individuals deeply invested in photography's empirical potential. These individuals were determined to expose the spirit photos for what they were. Not only were the images manipulated, according to this group, but that manipulation was *detectable*: manipulated images could be distinguished from un-retouched photos by *qualified experts*. According to Mnookin, these witnesses testifying *against* Mumler and his spirit photos but *in favor* of photographic evidence "invoked photographic manipulability not to dismantle photographic authority, but to preserve it."[38]

Hany Farid and others involved in developing computational techniques for digital image authentication share much in common with the nineteenth-century photographers who first advocated for photography's evidentiary worth. The computational techniques being developed to detect digital image tampering, while not foolproof, nevertheless promise to recover photography's claim to truth even in the face of digital imaging's seemingly unprecedented assault on photographic credibility. The intense interest in digital image manipulation and its detection – including the rise of a research program around this problem in the field of computer science – is proof-positive that the principle of photographic truth is still very much alive in the digital age. And the techniques being developed by twenty-first-century proponents of this principle are steeped in, and form a constitutive part of, the emerging epistemic virtue of computational objectivity.

Digital image authentication is only one of many techniques of forensic image analysis where we find evidence of an emergent investment in a computational form of objectivity. In fact, transforming raw video images into usable evidence itself usually requires altering image files, from simple enlargement of images or video frames, to more sophisticated techniques such as frame averaging. In other words, the process of forensic analysis typically involves manipulating images at the level of pixels and code. Image clarification techniques in particular can require analysts to modify images significantly from their original form. The fact that the production of usable evidence often *requires* the manipulation of images would seem to present an inherent challenge to the status of those images as evidence. But rather than entirely undermining the status of such evidence, this makes adherence to the code of computational objectivity that much more critical to the scientific and legal status of FVA (both the field at large and the results produced in any particular case).

Not only does FVA include many more techniques, but also the analysis of specific images is only the most obvious site where analysts must learn to think and see computationally. In any particular case, the need for computational thinking begins *before* any form of image analysis takes place, at the stage of field acquisition of video and the analysis of surveillance system configuration in site-specific contexts. The work of analyzing recorded video and producing valuable visual evidence from that video is preceded by the hardware and infrastructure inspection work needed to gather surveillance video from an almost limitless range of system configurations that investigators might encounter in the field. As Matthew Kirshenbaum has argued, in the digital age our "forensic imagination" – the desire to recover the past through inscriptions found in the present – requires a much better understanding of the materiality of electronic objects, the capacity to open up the black boxes of computer hard drives, for example, and examine how the storage and inscription of digital media happen.[39] The material dimensions of FVA are topic for another paper, but in any particular investigation involving digital video evidence, the encounter with the material forms of surveillance systems is precisely where the computational thinking begins.

Cops and computation

The legal and scientific legitimacy of FVA is in the process of negotiation, and it is inextricably connected to the professionalization of the field and its practitioners – so much so that legitimation and professionalization are in many ways the same process. The "forensic sensibility" of this new field is imbued with a sense of ethical responsibility to achieving what I am calling computational objectivity, drawing on Daston and Galison's history of objectivity. But despite the importance of this "epistemic virtue" to the work of FVA, computational techniques and computational thinking can only partially mitigate the problems associated with the forensic sciences. Strict adherence to computational objectivity can prevent neither the inadvertent introduction of bias nor outright falsification of evidence.

In fact, along with the effort to achieve computational objectivity, and to promote it as a new epistemic virtue, we find a renewal of the questionable promise associated with nineteenth-century forms of "mechanical objectivity" – namely, that the subjective and imperfect perceptual capacities of human beings can be eliminated from images and their interpretation. The belief in computational objectivity is a powerful counterweight to the challenge that digital techniques pose for the principle of photographic truth so important to the evidentiary value of images, whether in law, science, journalism, or other domains. But the belief that the subjective and imperfect interpretive capacities of human beings can be designed out of image analysis depends on a black-boxed view of both computational techniques and computational thinking. This view obscures the culturally specific models of human perception and interpretation that get programmed into the hardware and software of video analysis, as well as the myriad influences that shape the human labor and professional expertise that are central to the evidentiary production of video images.

In *Cop Knowledge: Police Power and Cultural Narrative in Twentieth-Century America*, Christopher Wilson examines the narrative authority the police claimed in the US throughout the twentieth century – how dominant ideas about crime and policing circulated back and forth from police policy and practice, to crime reporting and the popular genres of crime fiction and true crime storytelling.[40] As Wilson argues, police power is derived to a great extent from their narrative authority, or their ability to tell authoritative stories about crime and about their own proper role as arbiters of law and order. While struggles over legitimacy and authority have been fundamental to modern police institutions since their inception, what is changing today are the technologies and tactics on and through which these battles play out, and the forms of professional expertise required to make effective use of the evolving technological and tactical playing field. Wilson's critically important cultural analysis of policing needs updating for the digital age. What we find in the emerging field of FVA is a domain of cop-knowledge production and professional specialization, in a context of proliferating surveillance infrastructures and risk-management crime-control strategies. Although clearly fraught with the conflicting demands of science versus policing, the professionalization of FVA promises to invest the police with renewed narrative authority in the new-media landscape, allowing them to harness computational technologies and expertise in the constant battle of interpretations over the causes and consequences of crime, social disorder, and police power itself.

Notes

1 Clive Norris, Michael McCahill, and David Wood, "The Growth of CCTV: A Global Perspective on the International Diffusion of Video Surveillance in Publicly Accessible Space," *Surveillance and Society* 2, no. 2/3 (2004): 110–135, www.surveillance-and-society.org/articles2(2)/editorial.pdf.

2 On forms of citizen "cop watching," see Jesse H. Alderman, "Police Privacy in the iPhone Era? The Need for Safeguards in State Wiretapping Policing to Preserve the Civilian's Right to Record Public Police Activity," *First Amendment Law Review* 9 (Spring 2011): 487–542; Andrew Goldsmith, "Policing's New Visibility," *British Journal of Criminology* 50 (2010): 914–934; Benjamin J. Goold, "Public Area Surveillance and Police Work: The Impact of CCTV on Police Behavior and Autonomy," *Surveillance and Society* 1, no. 2 (2003): 191–203; Laura Huey, Kevin Walby, and Aaron Doyle, "Cop Watching in the Downtown Eastside: Exploring the Use of (Counter)Surveillance as a Tool of Resistance," in *Surveillance and Security: Technological Politics and Power in Everyday*

Life, ed. Torin Monahan (New York: Routledge, 2006), 149–166; and Dean Wilson and Tanya Serisier, "Video Activism and the Ambiguities of Counter-Surveillance," *Surveillance and Society* 8, no. 2 (2010): 166–180. In his analysis of "strategic incapacitation" as a new police strategy for managing political protest, Patrick Gillham notes: "Police agencies have changed substantially how they use the media in the era of strategic incapacitation by taking a 'proactive stance' to actively manage the flow of information to the media in the period leading up to, and during large protests as a way to channel the content of media coverage and deter public criticism" (p. 645). Patrick K. Gillham, "Securitizing America: Strategic Incapacitation and the Policing of Protest Since the 11 September 2001 Terrorist Attacks," *Sociology Compass* 5, no. 7 (2011): 636–652.

3 Goldsmith, "Policing's New Visibility," 916.

4 Richard V. Ericson and Kevin D. Haggerty, *Policing the Risk Society* (Toronto: University of Toronto Press, 1997).

5 Kelly Gates, "The Cultural Labor of Surveillance: Video Forensics, Computational Objectivity, and the Production of Visual Evidence," *Social Semiotics* 23, no. 2 (2013): 242–260, http://dx.doi.org/10.1080/10350330.2013.777593.

6 Thomas Levin, "Rhetoric of the Temporal Index: Surveillance Narration and the Cinema of 'Real Time'," in *CTRL [SPACE]: Rhetorics of Surveillance from Bentham to Big Brother*, ed. Thomas Levin, Ursula Frohne, and Peter Weibel (Cambridge, Mass.: MIT Press, 2002), 578–593.

7 Gates, "The Cultural Labor of Surveillance."

8 Jennifer Lee, "Caught on Tape, then Just Caught. Private Cameras Transform Police Work," *The New York Times*, May 22, 2005.

9 Quoted in Lee, "Caught on Tape, then Just Caught," p. A36.

10 Quoted in Avid, *Forensics Video Analysis Handbook: A Blueprint for Selecting a Forensic Video Analysis Workstation*, company brochure (Burbank, Calif.: Avid, 2004).

11 John Tagg, *The Burden of Representation: Essays on Photographies and Histories* (Minn. University of Minnesota Press, 1988), 75.

12 Tagg, *The Burden of Representation*, 75.

13 In the US, professionalization of policing is most commonly associated with the police-reform movement that took place during Theodore Roosevelt's tenure as New York City police commissioner in the 1890s: the institutionalization of formal academies, entrance exams, standardized uniforms, and systems for awarding merit certificates and medals. The period from the 1920s to the 1950s saw a "second wave" police-reform movement with an emphasis on rank-and-file training, bureaucratic efficiency, and autonomy from machine politics, as well as the rise of "police science" and the Uniform Crime Reports. Christopher Wilson, *Cop Knowledge: Police Power and Cultural Narrative in Twentieth-Century America* (Chicago: University of Chicago Press, 2000).

14 Allan Sekula, "The Body and the Archive," *October* 39 (Winter 1986): 3–64; Tagg, *The Burden of Representation*; Wilson, *Cop Knowledge*; Simon A. Cole, *Suspect Identities: A History of Fingerprinting and Criminal Identification* (Cambridge, Mass.: Harvard University Press, 2001).

15 Peter Manning, *The Technology of Policing: Crime Mapping, Information Technology, and the Rationality of Crime Control* (New York: New York University Press, 2008).

16 On police narrative authority see Wilson, *Cop Knowledge*.
17 LEVA, *Guidelines for Best Practice in the Forensic Analysis of Video Evidence*, www.leva.org/images/Best_Practices_FVA.pdf.
18 I took this course, "Level I: Forensic Video Analysis and the Law," from March 19–23, 2012, in the LEVA Digital Multimedia Processing Lab, located at the University of Indianapolis. From day one of the course, instructors placed considerable emphasis on memorizing the definition of FVA. Members of LEVA leadership themselves developed the definition, carefully wording it so that it could be presented in court in the process of establishing the professional credentials of expert witnesses.
19 National Academy of Sciences, *Strengthening Forensic Science in the United States: A Path Forward* (2009), *www.ncjrs.gov/pdffiles1/nij/grants/228091.pdf.*
20 See, for example, the essays published in Michael Lynch and Sheila Jasanoff, "Special Issue: Contested Identities: Science, Law, and Forensic Practice," *Social Studies of Science* 28, nos. 5/6 (1998): 675–686; and more recently Jay D. Aronson, *Genetic Witness: Science, Law, and Controversy in the Making of DNA Profiling* (New Brunswick, N.J.: Rutgers University Press, 2007).
21 Simon Cole, "Forensic Culture as Epistemic Culture: The Sociology of Forensic Science," *Studies in History and Philosophy of Biological and Biomedical Sciences* 44, no 1 (2013): 36–46.
22 *Melendez-Diaz v. Massachusetts*. 2009. 557 U.S. 305.
23 *Melendez-Diaz v. Massachusetts*. 2009. 557 U.S. 305.
24 *Melendez-Diaz v. Massachusetts*. 2009. 557 U.S. 305.
25 Wayne Cole, "New Burdens for Forensic Video Technicians," *Government Video* (September 4, 2009), www.oceansystems.com/documents/2009-09-New-Burdens-for-FVT-by-WCole.pdf.
26 Cole, "New Burdens for Forensic Video Technicians."
27 Cole, "New Burdens for Forensic Video Technicians."
28 Cole, "New Burdens for Forensic Video Technicians."
29 Cole, "New Burdens for Forensic Video Technicians."
30 Without a doubt, the push toward professionalization of forensic video analysts is coming to a significant extent from the courts. Both at the level of admissibility determinations and in the adversarial courtroom process, scientific objectivity is established in part through the professional credentials of expert witnesses. When surveillance video is introduced as evidence in court, a forensic video analyst is typically called to testify in order to establish the reliability of the process by which the particular video was obtained and made usable for the proceeding. In other words, an expert witness will often be called by the proponent of the evidence in order to establish that whatever was done to images to make them presentable as evidence was a legitimate image processing procedure. This approach to establishing the legal standing of photographic evidence is not all that different from how the courts have addressed admissibility since the earliest introduction of photographs in legal cases. In the US, the legal rules governing admissibility of visual evidence require judges to determine the *authenticity* of images, namely through the "pictorial testimony" or "silent witness" doctrines. The "pictorial testimony" rule requires a witness to testify that a piece of visual evidence is a fair and accurate representation of what it depicts. In contrast, the "silent witness" doctrine involves establishing the reliability of the image

production process. US courts have applied this doctrine to make determinations about the admissibility of digitally processed images. In other words, the proponent of the evidence attempts to establish the authenticity of digitally processed images with testimony from the person who processed the images, or by someone else capable of explaining the process.

Today a significant portion of video evidence is either born digital, or translated into digital form from an analog magnetic tape recording, and the reality is that much video evidence has been altered or enhanced in some way to make it suitable for analysis and presentation. To the extent that the process of FVA involves digital image enhancement or alteration, it invokes questions about the admissibility of digitally processed images – questions that the courts have had a number of occasions to consider over the last decade. The courts have tended to treat digitally processed images no differently than enlargements and other image-enhancement techniques that came before; however, it is nonetheless the case, as one legal analyst noted in 2007, that the "admissibility of enhanced digital images is an unsettled area of law in almost all jurisdictions." James M. Campbell, "Evidentiary Requirements for the Admission of Enhanced Digital Photographs," *Defense Council Journal* 74, no. 1 (2007): 12–21.

31 LEVA, *About LEVA*, www.leva.org/index.php/about.

32 LEVA, *Forensic Video Analysis Certification Program*, https://leva.org/index.php/certification; emphasis added.

33 Lorraine Daston and Peter Galison, "The Image of Objectivity," *Representations* 40 (Autumn 1992): 81–128.

34 Hany Farid, "Digital Image Forensics," *Scientific American* (June 2008): 66–71.

35 Farid, "Digital Image Forensics," 67.

36 Other conference panels and workshops included: "Digital Multimedia Workflow," focusing on the proper sequencing of work activities involved in video acquisition, processing, analysis, and presentation; and "Digital Multimedia Laboratory Best Practices," offering guidance on proper set-up of labs and management of lab activities. Additional topics addressed some of the more specific problems encountered in the forensic analysis of digital media, such as "Recovering Files from Damaged DVD or CD" and "Surviving Compression: What's Left in Images and Video." The conference included not only training workshops but also a "Digital Asset Management Expo," with exhibits by companies marketing surveillance and media forensic technologies, such as Reveal Media, which sells body-worn cameras, and MediaSolv, a company that bills itself as "evidence guardians" and sells digital evidence-processing systems. The professionalization of FVA is a characteristically neoliberal, public–private venture. Industry actors play key roles in the development of new technologies and forms of labor that define the emerging field of FVA and related domains of digital-media evidence production, analysis, and management. However, while the field of FVA is clearly dependent on private companies and their commercially available technologies, there is nevertheless recognition in the FVA community of the need to avoid being "sold" on unproven products. Comments of experienced analysts suggest that engaging in the work of FVA with professional integrity involves understanding the strengths and weaknesses of the various products, and having a healthy degree of skepticism about the promotional efforts of private companies.

37 Jennifer Mnookin, "The Image of Truth: Photographic Evidence and the Power of Analogy," *Yale Journal of Law and the Humanities* 10, no. 1 (1998): 1–74.
38 Mnookin, "The Image of Truth," 37.
39 Matthew Kirshenbaum, *Mechanisms: New Media and the Forensic Imagination* (Cambridge, Mass.: MIT Press, 2008).
40 Wilson, *Cop Knowledge.*

Chapter 4

Collision in a courtroom

Constance Penley

There is no richer site than a federal obscenity trial courtroom for observing the collision of images, ethics, and technology. This is especially the case in recent trials where judges have disallowed expert witnesses who can put the materials in context for the jury (in a pre-trial "Daubert" hearing) and declared that jury members do not need to see the entirety of the indicted materials to deliberate on them "taken as a whole." The defense is effectively deprived of the ability to mount a defense when the court decides in advance that sexually explicit adult materials have no context except themselves in all their sameness. In such a setting, who has the authority to determine which images can be seen and which cannot and what images mean and for whom?

When I was proffered as an expert witness in *United States v. John Stagliano, John Stagliano, Inc., and Evil Angel Productions, Inc.*, a federal obscenity trial that took place in July 2010 in the District of Columbia, I had no idea that I would be entering such a bizarre alternate universe where none of the logics of art, culture, technology, and everyday life as we know them apply.[1] But this is what happens when a judge instructs a jury to deliberate according to *Miller v. California* – the 1973 law that still governs obscenity prosecutions even in an era of the mainstreaming of porn and the explosion of accessibility on the Internet – and then proceeds to undermine any part of that law that might help to defend materials deemed not worthy of free-speech protection. In brief, the three-pronged "Miller Test" states that a jury can find that charged materials are obscene only if they find all three prongs to be the case:

1 that the average person applying contemporary community standards would find that the work, taken as a whole, appeals to the prurient interest;
2 that the average person applying contemporary community standards would find that the work depicts or describes, in a patently offensive way, sexual conduct or excretory functions specifically defined by applicable state law; and
3 that the work, taken as a whole, lacks serious literary, artistic, political, or scientific value.

Even though the 1973 decision introduced community standards and state law into a jury's deliberations, thereby opening up the possibility of many more local- and state-level obscenity prosecutions (and "forum shopping" for the most conservative communities and restrictive state laws), the third prong of Miller holds charged materials to a national standard that does not ask the jury to determine if they are utterly without redeeming social value (an earlier standard), but whether they are lacking in *serious* literary, artistic, political, or scientific value (the so-called LAPS test). Because juries must find that all three prongs apply for the material to be deemed obscene, and therefore illegal, not meriting First Amendment protections, and because juries may listen to a range of experts from around the country who can attest that the materials are not lacking in serious LAPS value, it is more difficult than one might think to get an obscenity conviction. But judges instruct jurors to follow *Miller v. California* and then whittle away at their ability to properly use it in their deliberations, especially prong three.

Let me first set up the scene of the trial. John Stagliano, one of the most successful and awarded adult directors and producers, and his company Evil Angel were charged in 2008 in a seven-count indictment with assorted violations of several federal obscenity statues. The government accused them of distributing in interstate commerce two allegedly obscene movies titled *Milk Nymphos*, *Storm Squirters 2: Target Practice*, and a movie trailer for *Fetish Fanatic 5* that was posted on a website allegedly owned by the defendant and his company. That last count about the Internet trailer for *Fetish Fanatic 5* was especially problematic because it got "dirtied up" by charging Stagliano with having distributed material that could be accessible to minors. What was the reasoning? A federal agent went into a hotel restaurant bar (a "public space"), opened his laptop and clicked on the trailer on the Evil Angel website. Logic would say that it was the federal agent who potentially exposed minors to *Fetish Fanatic 5*, though I do not think you find many minors in a bar.

To be clear, John Stagliano did not make any of these videos; they were produced by three members of his studio's stable of star directors: Jay Sin, Joey Silvera, and Belladonna. Stagliano's own considerable body of work, from his early hugely popular gonzo "Buttman" videos to his later high production value fetish films, such as the *Fashionistas* trilogy that was shot around the world and made into a Las Vegas musical, was not charged with obscenity. He was indicted for interstate commerce of obscene materials and, of course, pushing porn to children – always a boost to any prosecution of adult materials if you can bring in the children. The interstate commerce charge stemmed from an FBI sting of *Milk Nymphos* and *Storm Squirters 2*, where a federal agent at an office in Virginia ordered the videos online from a company in Maryland to be delivered to a PO box in Washington, DC. That Justice Department strategy of stinging the materials into the District of Columbia will become important later when we are looking at the extreme measures the court took to control the circulation of images in this trial.

Why was John Stagliano targeted for prosecution to the tune of 32 years in prison, a 7 million dollar fine, and the confiscation of his company and livelihood? Why these videos? And, for that matter, why the unusual decision to charge these materials for obscenity in a diverse and cosmopolitan region such as DC instead of, say, Alabama? Stagliano and Evil Angel were targeted in 2008 by the Bush administration's Obscenity Prosecution Task Force and would be one of only three cases held over into the Obama administration. Conservative lawmakers were outraged when Obama's Attorney General Eric Holder disbanded that task force in 2011, largely because of (spoiler alert) the government's prosecutorial ineptness in the Stagliano trial that resulted in a full acquittal, an embarrassing ineptness that could only be called *a catastrophically unethical mismanagement of images and image technologies at every phase of the trial.* Holder, who believes that such prosecutions are better handled by US Attorneys' offices and the Criminal Division's Child Exploitation and Obscenity Section – Democrats have tended to go after child porn, Republicans after adult porn – was attacked by a group of conservative lawmakers in 2011 who demanded the Department of Justice seek out the most mainstream producers in the adult industry, not just fringe producers such as Max Hardcore and Ira Isaacs. Those lawmakers believe if you can take out the mainstream adult films by the biggest and most respected directors then everyone else in the industry will be too scared to keep producing and the whole industry will be wiped out. These porn opponents want to deprive adult materials of their representational status (and any artistic claims or speech protections that might come with that status) and equate those films to prostitution, which is already illegal. After all, they claim, both the adult industry and prostitution are about paying someone to have sex – there's no difference between the two. But even some who support mainstream adult prosecutions concede that it is difficult when the products of this multibillion-dollar industry are readily available to anyone with a computer, and when the lines between pop culture and adult entertainment are irreparably blurred. It is likely for this reason that Holder has not begun any new prosecutions since he has been in office. But there is probably another reason why the Department of Justice targeted Stagliano, precisely for the mainstream status he had achieved in an earlier trial. In 2007, he won a large multimillion-dollar settlement against a Canadian company that was pirating his films (*John Stagliano, Inc. v. Kaytel Video Distributors*).[2] In that case he had an attorney, Allan Gelbard, who specialized in First Amendment law but also, and primarily, entertainment law and intellectual property. Gelbard put up two screens in the courtroom and played, side-by-side, Stagliano's films and the badly dubbed pirated prints; he won handily. Notably, the case had nothing to do with determining whether the films were obscene or not; it was a straightforward copyright infringement case, along with a right to publicity claim (the right to control the commercial exploitation of one's name, image, or persona), placing adult materials on the spectrum of all other creative materials to which one can establish ownership rights. The size of the award – at 11.2 million dollars, the largest ever received for copyright and trademark infringement by an adult entertainment company – showed the value of brands in the adult entertainment world. It also showed a court's willingness to defend that value. There is nothing anti-porn legislators or activist opponents fear more than having the adult industry perceived as just another cultural practice, just another business with valuable intellectual property and a good name to protect. Stagliano's winning of this case cemented his status as a legitimate major producer and creative artist, making him and his company, Evil Angel, a provocative target for those conservatives wanting to take down the most respected and mainstream of adult producers.

Why these videos? How did videos depicting women playfully exchanging bodily fluids with some light bondage thrown in become the most obscene thing in the land? After all, HBO's *Real Sex* has done specials on the subgenre of "women's play parties," where women engage in sexual exploration with each other and, if there's a man in the scene, he is just there to be a super-realistic live dildo. FBI Special Agent Dan Bradley, who was taken off national security issues and transferred to the obscenity desk for the trial, said he chose to order these videos because "[l]ooking at the covers [on the website], it appeared that those movies emphasized the excretory function and squirting."[3] In his opinion the jury would, of course, conclude that the average person in their community would find the videos patently offensive and appealing to prurient interest, and that no qualified expert could offer credible evidence that the videos were not lacking serious value. Maybe the government was hoping that so many instances of female ejaculation in these squirting videos, including big white ejaculations of milk out of the ass in *Milk Nymphos* and some spectacular Olympic-level ejaculations in *Storm Squirters 2: Target Practice*, would be so shockingly repulsive to the jury that they would immediately declare them to be obscene. Prosecutors in obscenity cases believe that all they have to do is simply show the movies to the jurors and say, "I rest my case."

Why the District of Columbia? Most obscenity prosecutions are mounted in the most religious and conservative parts of the country where you can be sure of getting the verdict you want. It seems that the Evil Angel videos were originally going to be prosecuted in reliable Alabama, but then strangely moved to the much more cosmopolitan District of Columbia, which had not seen an obscenity trial in almost a quarter of a century. Could it have been moved from Alabama because US Attorneys there were resistant to the Karl Rove-inspired politicization of their jobs in the Alberto Gonzales Justice Department? Of the nine US Attorneys Gonzales fired, one of the most prominent was the Nevada state attorney who was not considered aggressive enough in prosecuting (Democratic) voter fraud and pornography.

Or, it may have been that the very conservative judge, Richard Leon in the DC court where the case landed, wanted a splashy obscenity trial under his belt on the way to a Supreme Court nomination. (Leon had devotedly worked for Republicans on everything from the October Surprise to Iran/Contra to Whitewater.) This is speculation because there was so little transparency to the proceedings and everything that led up to them, a lack that became a hallmark of the trial.

Judge Leon declared that the indicted movies did not need to be shown in full at trial.[4] This poses a serious problem for the defense, because the three prongs of the Miller Test, which is the backbone of jury instruction under current obscenity law, explicitly require that jurors evaluate and make decisions on a wide range of issues (including artistic intention and merit), which can only be determined by viewing films in their entirety. The judge proceeded to make ruling after ruling that limited and controlled how the jury, the press, and the public would have access to the materials on trial. For example, Judge Leon directed that the video monitors face the jury so that no one else in the courtroom could see the images. And only the jurors, wearing headphones, could hear the soundtrack. Such decisions deprived the media and public of crucial information for evaluating the fairness and openness of the trial. So, too, these decisions, made in the interest of decorum in the courtroom, automatically stigmatized the images as unfit for public consumption.[5]

Before getting back to the machinations of the trial, let me say briefly how and why I was chosen to be one of the two expert witnesses for the defense in *United States v. John Stagliano*. (The other expert witness was Dr. Lawrence Sank, a clinical psychologist in the Greater Washington area who uses sexually explicit fetish materials such as the charged videos in his therapeutic practice and was expected to speak to their therapeutic and scientific value.) I have been teaching a course on pornographic film as a genre and an industry, as film and popular culture, at the University of California, Santa Barbara, since 1993. In first broaching the course to my colleagues, I reminded them of our recent external review that called our undergraduate program "likely the best in the country." We were praised for the depth and breadth of our curriculum across authors, national cinemas, history, theory, directors, and genres. But, I said: "We are supposed to be the best, the most comprehensive, but we are not teaching the most enduring and prolific of all film genres." They said: "Oh, you are right; of course." I also pointed out to them that we could do a better job as University of California professors producing research leading to an understanding of the culture and economy of the state of California if we studied the center of the adult film industry, just 60 miles to the south of us in the San Fernando Valley. So, to the course listings under the rubric of "topics in film genre," where the professor fills in the western, the musical, science fiction, or horror, I inserted pornography. My colleagues wanted to know how I was going to put together a teaching collection as exhaustive as the collections for teaching our other genre and history classes. I told them that I was able to amass a teaching collection from the Institute for the Advanced Study of Human Sexuality (IASHS) in San Francisco. The IASHS archivists were so delighted that someone was going to teach pornography as film. They had often supplied films to professors over the decades as it turned out, for courses on human sexuality or law and society, or a graduate seminar on representation and sexuality, but those professors would typically show only one hardcore film as an example of that aspect of sexuality. I wanted to show pornographic films over the entire course. My earliest film is from 1907 and I am able to go up to the latest big-budget porn parodies and, of course, the explosion of porn on the Internet and for mobile technologies. For the most part, my students are advanced film and media majors, because I want them to be able to ask of pornographic film and media all the questions they normally ask of all other forms of film, including Hollywood film. How have the style, strategies, and content changed over the decades? What have been the modes of production, consumption, and distribution? What have been the venues and audiences? Has porn led technology or been led by it? How has the legal climate in any given era shaped all of the above?

What I want my students to be able to do is something that not many people have the chance to do. As a society, we debate, legislate, and regulate pornography in an almost total vacuum of knowledge about what it really consists of historically, textually, and institutionally. My students are welcome to their opinions, but they can speak about pornography only after they have had the far too rare opportunity of being able to learn what it is.

Because in Santa Barbara we are so close to the industry, I am able to bring to my class the leading lights and working stiffs of the adult industry. These include performers, directors, studio heads, members of the industry trade organization, publishers of the leading industry trade journals, attorneys, and representatives from the now defunct Adult Industry Medical Healthcare Foundation. One of my most regular guests has been John Stagliano, who was planning to be my guest anyway when his videos were first charged in 2008. I said, John, bring the videos to my class; showing them in a Research 1 institution will give them scientific value. So we looked at them in 2008. My students, who cannot get their heads around obscenity law, and who by this time had seen huge amounts of porn for comparison, could not believe that these videos featuring female ejaculation had been declared the most obscene thing in the land. In 2010, just before the trial, Stagliano came back to my class with the videos and his personal attorney, Allan Gelbard, who had won the 2007 multimillion-dollar copyright infringement case for him.

One of the things that made my experience with the trial so fascinating was being able to hang out with Stagliano's dream defense team, including two of Larry Flynt's attorneys, Paul Cambria, and Robert Corn-Revere; the attorney who successfully defended Dennis Barrie, the head of the

Cincinnati Contemporary Art Center charged with obscenity for his Robert Mapplethorpe show, Louis Sirkin; and the attorney who got the posthumous pardon for Lenny Bruce, again Robert Corn-Revere. But Allan Gelbard was not one of these lions of the First Amendment. He was younger and more technologically savvy than the other attorneys (and had been a pornographer himself). His greater knowledge about how images and technologies get deployed in the courtroom would prove crucial to the trial. But it was Allan Gelbard who saw how porn was presented in my class as film and popular culture, who thought that the indicted materials could be defended not just on the grounds of high art – Art with a capital A – but also popular culture. He had to convince the other attorneys, who were very wed to what had so often worked for them – that porn was "Art" and therefore meriting the protection of free speech – to choose me.

An abiding theme of my research has been the importance of popular culture as a site of crucial conversations about who we are and what kind of world we want to live in. Some of you may know my work on the underground female media fans (the "slashers") who since the 1970s have taken the male characters and fictional universes of popular television shows to create pornographic, homoerotic stories to exchange as gifts among other women writers. They have written their desires for better men – men who aren't afraid to share their emotions, men whose personalities aren't all twisted by homophobia – and their wish for a different and better, more egalitarian world in zines, novels, poems, art, and video. And in other research that I have done on pornography as male popular culture, in a piece called "Crackers and Whackers: the White Trashing of Porn," I argue that the kind of lewd humor seen in everything from early stag films to *John Wayne Bobbit: Uncut*, from Howard Stern to *Beavis and Butthead* and *South Park* all contain critical conversations about masculinity, where it is more often than not the man who is the butt of the joke for his sexual and social ignorance, often at the hands of trickster women.[6] I conclude that if this critical conversation among men is happening through the popular culture targeted to men – films made for men by men – then that is important speech that should be of great interest to feminists and it should be protected. I walked the feminist plank on that one but, surprisingly, few have criticized my defense of pornography as male popular culture.

So, now, try to imagine taking this kind of argument to a federal obscenity courtroom that will only recognize "serious" artistic value and where the judge, not the jury, gets to determine what that might be and whether the proffered expert witness has got the science on her side to prove it. Something we never got to in defending these materials, because I was excluded and then the whole case fell apart because of the government's ineptness, was arguing for these videos as not only not lacking in serious artistic value but also serious political value. I did not want the Justice Department to see my syllabi for the class, to see the way I teach porn, because they might be able to know in advance that I would be defending porn for its political (feminist) value as well as its artistic (as "Art" and as the art of popular culture) value. I also thought the game would be over if they had read any of my writing on women, porn, art, and popular culture, alongside my writing on the value of pornography as male popular culture. But why did I think they could read?! The other reason why I did not want them to see my syllabi (other than not liking the idea of the government scrutinizing my teaching materials) was that they would see that John Stagliano was a frequent guest lecturer in my class and that we had screened and discussed the indicted materials twice. Fortunately, what they could not see was that my students had taken on the mighty job of transcribing all the videos to help me in my expert testimony, briefly turning my class into the Innocence Project for Porn. So, too, if the prosecutors saw that Stagliano had been to my class so often, they could have accused me of testifying for him because he was my friend. I was looking forward to being able to say on the stand: "The first time I saw Mr. Stagliano outside of my class was for this trial." I was also looking forward to saying in answer to the inevitable discrediting question to an expert witness – "How much were you paid for your testimony today?" – that I was charging nothing, that I was doing it for my research. So even though they got the syllabi from me by subpoenaing them, I do not think they learned anything from them (Figure 4.1). Their mindset is just so different.

With that background, then, I was proffered as the expert witness and had to be qualified as such in a pre-trial Daubert hearing where the judge gets to determine whether the testimony of the proffered expert witness is scientifically credible and will aid the trier of fact – that is, the jury – to understand the evidence or to determine a fact in issue. According to the Federal Rules of Evidence,

> a witness qualified as an expert by knowledge, skill, experience, training, or education, may testify in the form of an opinion or otherwise [that is, my opinion is that these materials do not lack serious artistic value] if (1) the testimony is based upon sufficient facts or data, (2) the testimony is the product of reliable principles and methods, and (3) the witness has applied the principles and methods reliably to the facts of the case.[7]

The law that mandates a Daubert hearing has been called "the most influential Supreme Court decision that you've never heard of."[8] The Daubert rule came out of the Supreme Court's decision in *Daubert v. Merrell Dow Pharmaceuticals, Inc.* (1993) and in related cases involving General Electric (GE) and a tire company that established a framework to guide trial courts in performing their "gatekeeping" function. Under Daubert, if a party proffers expert testimony that is scientific in nature, it is admissible only if the trial court (that is, the judge) concludes: (1) that the reasoning or methodology underlying the testimony is scientifically valid, and (2) that

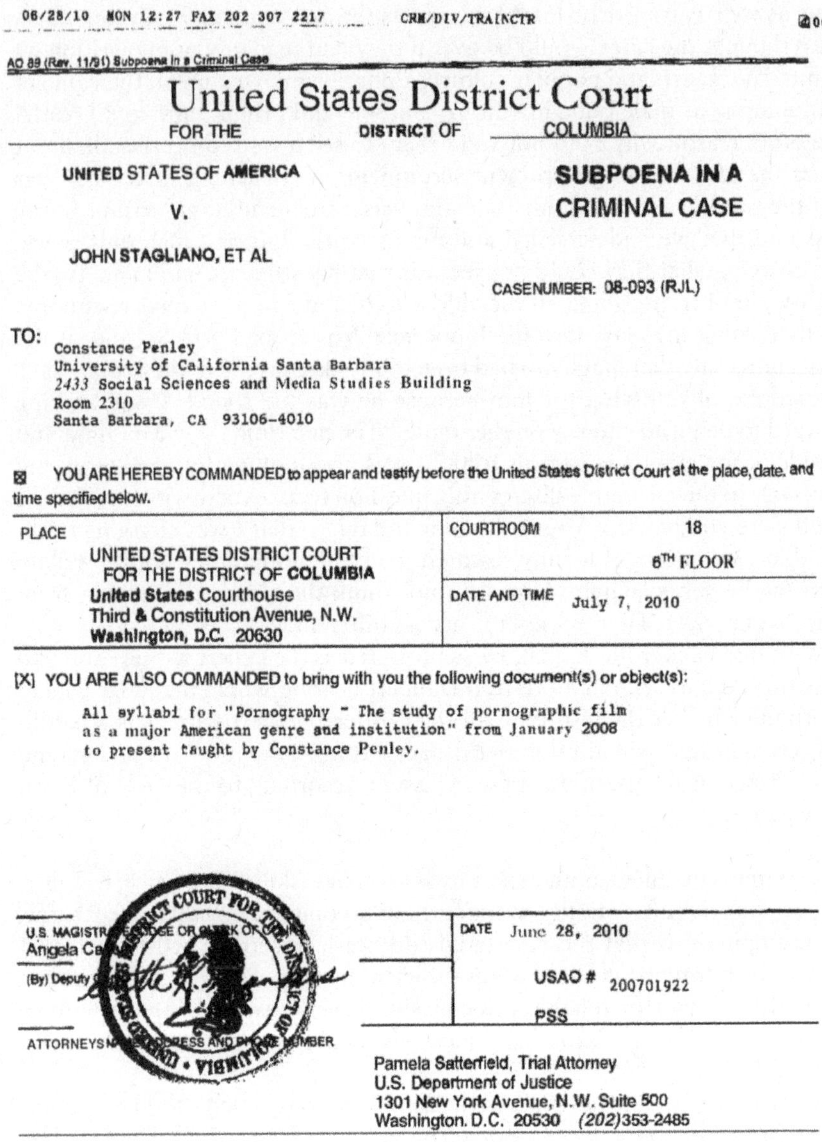

06/28/10 MON 12:27 FAX 202 307 2217 CRM/DIV/TRAINCTR ☒002

AO 89 (Rev. 11/91) Subpoena in a Criminal Case

United States District Court

FOR THE DISTRICT OF COLUMBIA

UNITED STATES OF AMERICA

V.

JOHN STAGLIANO, ET AL

SUBPOENA IN A CRIMINAL CASE

CASENUMBER: 08-093 (RJL)

TO: Constance Penley
University of California Santa Barbara
2433 Social Sciences and Media Studies Building
Room 2310
Santa Barbara, CA 93106-4010

☒ YOU ARE HEREBY COMMANDED to appear and testify before the United States District Court at the place, date, and time specified below.

PLACE	COURTROOM 18 6TH FLOOR
UNITED STATES DISTRICT COURT FOR THE DISTRICT OF COLUMBIA United States Courthouse Third & Constitution Avenue, N.W. Washington, D.C. 20630	DATE AND TIME July 7, 2010

[X] YOU ARE ALSO COMMANDED to bring with you the following document(s) or object(s):

All syllabi for "Pornography - The study of pornographic film as a major American genre and institution" from January 2008 to present taught by Constance Penley.

U.S. MAGISTRATE JUDGE OR CLERK OF COURT
Angela Ca[...]
(By) Deputy Clerk

DATE June 28, 2010

USAO # 200701922

PSS

ATTORNEY'S NAME, ADDRESS AND PHONE NUMBER

Pamela Satterfield, Trial Attorney
U.S. Department of Justice
1301 New York Avenue, N.W. Suite 500
Washington. D.C. 20530 (202)353-2485

Figure 4.1 Penley syllabi subpoena from *United States v. John Stagliano.*

Matthew Bowman

From:	STEVEN TAKESHITA [22202@lapd.lacity.org]
Sent:	Monday, June 28,2010 9:46 AM
To:	Matthew Bowman
Cc:	PATRICK SHIELDS; KYLE LEWISON; MICHAEL OZAKI; Pamela Satterfield DOJ DC Attny
Subject:	Subpoena
Attachments:	Penley Subpoena.pdf

Sgt. Bowman,

Per our telephone conversation, I am emailing you the federal subpoena for Prof. Dr. Constance Penley. Please confirm the service of the subpoena and/or any difficulties with the service.

Please have Dr. Penley contact US DOJ Attorney Pamela Satterfield for instructions about the trial, items to be brought with her and the dates Dr. Penley will be needed for trial. Pamela Satterfield's contact information is listed on the bottom of the subpoena.

If you have any further questions, please feel free to contact me at the below listed numbers.

Thank you for you and your agencies cooperation,

Detective - III
Steven M. Takeshita 22202
Officer-In-Charge
Pornography Unit
Community Problems Unit
Electronics Unit
Detective Support and Vice Division
Los Angeles Police Department
100 W. 1st Street #441
Los Angeles CA 90012
(213) 486-0910
(213) 486-0909 Fax

CONFIDENTIALITY NOTICE: The information in this transmittal is personal **& confidential.** It is intended only for the **recipient(s)** listed above. Any review, use, disclosure, distribution or copying of this transmittal is prohibited except by or on **behalf of** the intended recipient. If you have received this **transmittal** in error, please notify me immediately by reply **email & destroy** all copies of the transmittal.

Figure 4.1 (continued)

the reasoning or methodology will assist the trier of fact to understand or determine a fact in issue. The focus is "solely on principles and methodology, not on the conclusions that they generate." To determine whether a particular theory or methodology is "scientifically valid," courts generally look to several factors: (1) "whether the theory or technique can be and has been tested"; (2) "whether the theory or technique has been subjected to peer review and publication"; (3) "the method's known or potential rate of error"; and (4) "whether the theory or technique finds general acceptance in the relevant scientific community." These factors are not exhaustive, nor are they dispositive or even applicable in every case. To be sure, the inquiry is a flexible one, but the overarching purpose is to determine whether the proffered testimony is reliable – that is, grounded in the scientific method – and, if so, whether it is relevant or helpful to the jury's factual inquiry.

But the Daubert framework is not limited to scientific testimony. It is equally applicable to any testimony that is based on technical or specialized knowledge. Particularly persuasive is the advisory committee's note for Rule 702, which explains that "[w]hile the relevant factors for determining reliability will vary from expertise to expertise, [Rule 702] rejects the premise that an expert's testimony should be treated more permissively simply because it is outside the realm of science." Indeed, on its face, Rule 702 draws no distinction between the requirements for qualifying opinion testimony based on scientific knowledge and the requirements for qualifying opinion testimony based on any other specialized knowledge. A lot of this language here is Judge Richard Leon's set-up language in his decision that eliminated me as an expert witness, along with clinical psychologist, Dr. Lawrence Sank.

Where did this Daubert framework come from? I mentioned trials involving Dow, GE, and a tire company. It came from those corporations' efforts to get rid of what they called "junk science" as admissible evidence. So, if expert witnesses for members of a community are complaining about all the toxic materials that have been leached into their soil and water or a physician who says eight teenagers in his care committed suicide from taking that company's antidepressant, the corporations can throw all of their scientists at the other side's experts. This used to be done in front of a jury. After Daubert, the judge decides what is admissible because it is credibly scientific – the *judge*, who may have no training in chemistry, psychiatry, or, in my case, film and media studies. Here is Judge Richard Leon: "On July 9, 2010, in a ruling from the bench, I rejected the parties' proffered expert witnesses because the defendants had failed to establish that the principles and methodologies underlying their proposed expert testimony were sufficiently reliable and useful to the jury to satisfy Federal Rule of Evidence 702."[9]

In the Daubert hearing, the defense presented my credentials, my academic training, and any other professional service I had done on eroticism in art. The prosecution had the job of questioning me to prove that my credentials were not valid and that the testimony based on my "expertise" would have no usefulness to the jury in its deliberations. Judge Leon, who had already let it be known that he was going to be hostile to the idea of expert witnesses at all, had been giving the defense team a horrible time for two years, dismissing every motion and letting them know that nothing was going to go their way, turned out to be a film buff. The defense team told me after my testimony that this had been the best day they'd had in two years. When I was describing how I teach pornography as a film genre, Judge Leon leaned in eagerly and said: "Like film noir?" I was not going to bust his film cred by telling him film scholars do not consider it a discrete genre but more like a thematic element. I just said: "Oh, yes." In the part of the Daubert hearing where Allan Gelbard was leading me through my credentials, he asked, because I had told him about this: "Have you ever done any professional work outside of academia that might inform your testimony?" I talked about the extensive research project I did for Francis Ford Coppola on eroticism in classical film, in relation to his film *One from the Heart*. Judge Leon was beaming. I think he was a Coppola fan. When the deputy prosecutor was challenging my credentials, the only thing she could come up with was that my teaching collection was not academically legitimate because it came from the Institute for the Advanced Study of Human Sexuality – not in fact an accredited PhD granting program outside the state of California – the judge called a sidebar. I found out afterwards that he told the deputy prosecutor to give it up already because my CV was 25 pages long.

So why did I get excluded, when, as the defense team said: "He loved you, he was all over you!" One thing that I have learned in obscenity trials is that what you see in the courtroom has almost nothing to do with what is really going on. Judge Leon had to write a brief excluding me and the other proffered witness for something that was completely behind the scenes, which I cannot really talk about just yet. In his tortured and tortuous decision, he twisted every which way to say that while my credentials were impressive, I had never written or lectured about why pornography was "Art" and I had insisted that taste could play a role in our judgments, which meant that my views were merely subjective. I had also refused under the deputy prosecutor's questioning to say whether, using the analytical conventions of my discipline, I thought this film or that film would be art or pornography. If it had such and such sex acts in it, would it be art or pornography? I kept saying that the whole point of my research and teaching was that you had to *see it* before you could make statements or offer opinions about it. When she asked me that question about a film that included such and such acts and involved children, I said, again, I could not say without seeing it and reminded her that it would be illegal for me to see it and illegal for her to show it to me.

Even if my expert testimony had not been excluded, the defense would not have had to use me because the government's case turned into a technological-ethical debacle that led to Stagliano and his company being completely acquitted of all charges.

The first charge to go was the count saying that Stagliano had made materials that could be accessible to minors. When the government, on screens tilted toward the jury so that no one else in the courtroom could see the images (already prejudicing the jury about the horribleness of the images), played the disc of the downloaded *Fetish Fanatic 5* trailer from the Evil Angel website, at first there was no sound. Then when they got the sound to, sort of, play, the image started freezing and glitching up. It seems the disc was corrupted, which Allan Gelbard had warned the Justice Department about two years earlier in the discovery phase. You would think the government would have tested the audiovisual set-up for this crucial piece of evidence, but it appears not. Even Judge Leon, who had said he did not think the jury needed to see all the charged material to make its determination, was discomfited when the FBI agent could not say whether that glitched-out, corrupted disc was indeed an exact copy of the trailer he had downloaded from the Internet. The child-porn charge was thrown out.

But Judge Leon was still fine with allowing the prosecution to show only select parts of the charged videos and for the FBI agent to attest that the content of the rest of the video was essentially the same. What scene from *Milk Nymphos*, for example, did they choose for this largely black and female Washington, DC jury? Scene number two where a black milkman delivers milk to two white women who do lots of things with that creamy stuff while talking constantly about his big "n-word" cock. Again, we will never know whether that would have swayed the jury (though some media interviews with jury members afterwards suggested that it had not), because the prosecution also could not prove that the videos had ever been viewed by the agent in the District of Columbia where the jury was supposed to deliberate on these videos based on their local community standards. The FBI agent who had stung the videos into that DC PO box sent another agent from their branch office in Maryland to pick them up and bring them to the office in Virginia. But all of this corruption of the chain of evidence was compounded when the prosecutor asked the mandatory question of the FBI agent: "When was the last time you reviewed the videos?" The agent answered that he had looked at them a couple of weeks ago when the prosecutor told him the judge wanted to be sure his memory of the videos was sharp for his testimony. Everything came crashing to a halt. The judge had just been accused of helping to coach the star witness. The prosecutor, to save herself, would have to say that the agent lied, that the judge never told her that, and she had never passed the judge's caution along to the agent. But you have to recuse yourself if you charge your own star witness with lying and, if the prosecutor did that, the case was over.

At this point everything about the intersection of images, ethics, and technology in this courtroom had been corrupted. The defense called for a Rule 29 motion. After the government closes its evidence (or after the close of all evidence), the court on the defendant's motion must enter a judgment of acquittal of any offense for which the evidence is insufficient to sustain a conviction. That judge just wanted to get out of that courtroom as quickly as possible and almost immediately granted the motion to acquit and excoriated the government for doing only a "good enough for government" job. Though Stagliano and his dream defense team were happy that he was not going to prison for what could have been the rest of his life, they were disappointed that this case, with so many appealable errors – including disqualification of expert testimony and depriving the jurors of the chance to consider the charged materials "taken as a whole" – would not be going to the Supreme Court, with the chance that some provisions of our archaic obscenity law would be re-decided.

My own biggest disappointment that the trial concluded with an acquittal and the defense never got to make any of its arguments is this: at one moment during the defense, the plan was to ask if the FBI agent who had stung the materials into DC knew that *Milk Nymphos*, *Storm Squirters 2*, and *Fetish Fanatic 5* were already in DC, at the Library of Congress where they had been duly deposited for copyright registration.

Notes

1 *United States of America v. John Stagliano, John Stagliano, Inc., and Evil Angel Productions, Inc.* United States District Court for the District of Columbia, Criminal Action No. 08-93 (RJL), filed April 4, 2008, https://ecf.dcd.uscourts.gov/doc1/04512043338 (retrievable through Public Access to Court Electronic Records, www.pacer.gov).

2 See, for background on *John Stagliano, Inc. v. Kaytel Video Distributors (Central District of California, 2007)*, James R. Alexander, "Swarms and Trolls – Recent Trends in Copyright Protection of Commercial Pornography in the United States," *Social Science Research Network* (March 30, 2013), http://ssrn.com/abstract=2260913. For a historical overview on pornography and copyright, see James R. Alexander, "Evil Angel Eulogy: Reflections on the Passing of the Obscenity Defense in Copyright," *Journal of Intellectual Property Law* 20 (spring 2013): 209–313.

3 Amanda Hess, "Buttman v. The Man: D.C.'s First Big Obscenity Trial in Decades Fails to Determine the Obscenity of Milk Enemas," *The Washington City Paper*, July 22, 2010, www.washingtoncitypaper.com/blogs/sexist/2010/07/22/buttman-v-the-man-dcs-first-obscenity-trial-in-decades-fails-to-determine-the-obscenity-of-milk-enemas/.

4 See Judge Richard Leon's attempted refutation of the requirement under *Miller v. California* that the jury make their deliberations on the charged materials, "taken as a whole": *United States of America v. John Stagliano et al.*, Criminal Action No. 08-93 (RJL) 693 F. Supp. 2nd 25; 2010 U.S. Dist. Lexis 14770 (February 19, 2010) B.

5 Senior Legal Editor at AVN Media Network, Mark Kernes, provided the most comprehensive coverage of the trial. He wrote a letter to the court to object to the lack of press access to much of the trial proceedings. "AVN Reporter Airs Concerns on Stagliano Case in Letter to Court," *AVN*, July 13, 2010, http://business.avn.com/articles/legal/AVN-Reporter-Airs-Concerns-on-Stagliano-Case-in-Letter-to-Court-403151.html. The trial received very little mainstream-media coverage (a *Washington Post* blogger told me, "it's not my paper's brand"). But Las Vegas-based journalist Richard Abowitz covered it very thoroughly for *Reason*, which was a big supporter of the libertarian John Stagliano, who has been a donor to both the Reason Foundation and the Cato Institute. See Abowitz's excellent summary article "The Trial of John Stagliano," *Reason Foundation*, July 12, 2010), http://reason.org/news/show/trial-john-stagliano.

6 Constance Penley, "Crackers and Whackers: The White Trashing of Porn," in *Porn Studies*, ed. Linda Williams (Durham, N.C.: Duke University Press, 2004), 310–331.

7 "Rule 702: Testimony of an Expert Witness," *Federal Review of Evidence* 30 (2014), http://federalevidence.com/downloads/rules.of.evidence.pdf.

8 "The Most Influential Supreme Court Decision You've Never Heard of," *Defending Science.org*, www.defendingscience.org/daubert-most-influential-supreme-court-decision-youve-never-heard.

9 Memorandum Opinion, *United States v. John Stagliano et al.* July 30, 2010: 2, http://federalevidence.com/downloads/rules.of.evidence.pdf (retrievable through Public Access to Court Electronic Records, www.pacer.gov).

Chapter 5

"Who speaks for the art?"

Larry Gross

The documentary, *The Art of the Steal*,[1] is an unabashedly tendentious account of the successful campaign to move the Barnes Foundation from Merion, Pennsylvania to downtown Philadelphia, a move the film describes as "[t]he greatest act of cultural vandalism since World War II."[2]

Towards the end of the documentary, journalist John Anderson, author of an earlier account of "The Battle Over the Barnes collection," *Art Held Hostage*, says: "The forces that in effect are keeping the Barnes hostage are almost overwhelming. You could ask the simple question: Who speaks for the art, or the legacy of Dr. Barnes, when so many powerful political and economic forces are at work against it?"

Another talking head in the documentary is Julian Bond, then chairman of the National Association for the Advancement of Colored People, and introduced in the film as the son of Horace Mann Bond, who had been president of Lincoln University, a small historically black university outside Philadelphia, in 1950 when Dr. Albert Barnes amended the Barnes Foundation indenture to specify that "the next four vacancies [on the Board of Trustees] . . . shall be filled by election of persons nominated by Lincoln University, Chester County, Pennsylvania,"[3] giving effective control of the Barnes to Lincoln University.

Among the final statements in the documentary is Julian Bond's summary assessment that "[Barnes] will be violated in the experience he wanted you to have. And that's important, because it was *his* art. He had the right to do with it as he chose. And these people – these vandals – stepped in and took it away from him."

There are many interesting artistic, ethical, legal, and political questions raised by the film and, indeed, by the transfer of the Barnes Foundation. I want to focus here on a few of these questions, in the context of a consideration of the peculiar role and standing of the collector, the owner, of works of art.

Note Anderson's phrasing: "Who speaks for the art, or the legacy of Dr. Barnes, when so many powerful political and economic forces are at work against it?" In a subtle turn of phrase he is equating the "legacy of

Dr. Barnes" with the interests of the art, whatever they might be. Should we grant that respecting Dr. Barnes' legacy is the ethical equivalent of "speaking for the art"? After all, wouldn't it be fair to assume that the interests of the artists are relevant when it comes to asking "who speaks for the art"?

Pushing this a bit further, Julian Bond asserts that Barnes "will be violated in the experience he wanted you to have" he is on reasonably firm ground, although that, too, is somewhat debatable, once we note that the new Barnes on the parkway faithfully re-creates the interior design and idiosyncratic layout of the Merion original, albeit in an altogether new setting and wrapped around by a new exterior. But, what about Bond's next sentences: "And that's important, because it was *his* art. He had the right to do with it as he chose. And these people – these vandals – stepped in and took it away from him"?

Surely, we can presume that Bond doesn't literally mean that Barnes had the moral – as opposed to the legal – right "to do with it as he chose." To make the point by absurd extension, I trust that Bond would not assert that Barnes had the "right" to consign the paintings to a bonfire were he to grow tired of them, or decide that they were not worthy of preservation. Legal right, yes; although the artists in question, and likely Dr. Barnes, would have been familiar with French law of *droit moral*, which since the 1890s has recognized interests of artists in their work.[4] These rights are enshrined in the 1928 Berne Convention for the Protection of Literary and Artistic Works: "Independently of the author's economic rights, and even after the transfer of said rights, the author shall have the right to claim authorship of the work and to object to any distortion, mutilation, or other modification of, or other derogatory action in relation to, the said work, which shall be prejudicial to his honour or reputation."

At the time of Barnes' death in 1951 there were no such legal protections for artists under US law. In 1990, following the requirements of the Berne Convention, the US instituted the Visual Arts Rights Act (VARA). VARA was the first federal copyright legislation to grant protection to *moral rights*. Under VARA, a painter may insist on proper attribution of his painting and, in some instances, may sue the owner of the physical painting for destroying the painting even if the owner of the painting lawfully owned it.

So, let's assume that Julian Bond is not endorsing Barnes' right to destroy the paintings. We might also ask, though, about the moral right to hide the art away, say in a bank vault, or even have it cremated along with his body, as Japanese financier Rayoe Saito is reputed to have planned to do with Van Gogh's *Portrait of Dr. Gachet* and Renoir's *Au Moulin de la Galette*.[5] In fact, the Pennsylvania Supreme Court, in ruling that the Barnes Foundation had to become more accommodating to the public, noted: "A work of art entombed beyond every conceivable hope of exhumation would be as valueless as one completely consumed by fire. Thus, if the paintings here involved may not be seen, they may as well not exist."[6]

While we are considering the ethical as well as the legal rights of those who happen to own or otherwise control works of art, let us briefly note some important cases that demonstrate the complexities of these matters:

- Alexander Calder was unsuccessful in his efforts to force the County of Pittsburgh to reverse its actions in rendering his mobile, *Pittsburgh*, which had been donated to the Pittsburgh airport, immobile, and painting it in the county colors of green and gold. Only years later, after his death, did national publicity persuade them to undo the modifications.[7]
- The curator of the Wichita State Art Museum allegedly painted some of Calder's unpainted works to increase their market value, in order to please the donor, who wanted a larger tax deduction for donating the works, and who knew that painted works would receive a higher appraisal.[8]
- After the death of sculptor David Smith it became known that influential critic Clement Greenberg, who had been a mentor and booster of Smith, authorized changes in several of Smith's sculptures. Greenberg had objected to Smith's decision to apply paint to his metal sculptures, even though Smith himself had written a letter "renouncing" one of his works after what he termed "a willful act of vandalism" that caused the paint to be removed.[9] Greenberg, acting as executor, "ordered the removal of white paint from a number of Smith's painted sculptures in the open air so that their painted surfaces would be eroded by the natural effects of the weather."[10]
- When Franz Kafka died in 1924, he left instructions to his close friend, Max Brod: "Dearest Max, My last request: Everything I leave behind me . . . in the way of diaries, manuscripts, letters (my own and others'), sketches and so on, to be burned unread." As we know, Brod decided to disobey the writer's wishes, and proceeded to publish *The Trial*, *The Castle*, and *Amerika*, as well as a biography of Kafka. When Brod fled to Palestine in 1939, he took Kafka's papers with him, and, during the next few decades, he edited and published more of Kafka's writings.[11]

I believe it's fair to presume that most of us would find the Pittsburgh authorities philistine, the Wichita curator venal, and Greenberg arrogant. But it's also likely that we are grateful that Max Brod disregarded his friend's explicit wishes.

So, let's pull back from the extreme and look at the narrower claim, that Barnes "will be violated in the experience he wanted you to have." Here, we have to travel back a century or more, to see what sort of experience Barnes was aiming at. We need, in a sense, to bracket our own experiences with art museums, to see things in a different light.

Museums begin to emerge in the eighteenth century as collections such as those of the Vatican, the Sloane Collection in London that became the

British Museum, and the royal collections in Vienna are opened to the public. Following the French Revolution the Louvre was opened in 1793, and the dislocation of art during the revolution and the Napoleonic wars scattered numerous private collections that were later reassembled as national collections.[12] These new institutions are constituted as the property of the public. "It is for the public that society in the new democratic age retraces in social space – through the creation of zoos, libraries, parks, museums, and concert halls – the amenities of leisure and privilege once held by a few within the private space of moneyed or aristocratic property."[13]

The public that these museums addressed, however, were those sufficiently educated and affluent to appreciate the works in their new context, as works of art, whatever they might have been in their original context. "It would be an act of madness to enter a museum, kneel down before a painting of the virgin to pray for a soldier missing in battle, lighting a candle and leaving an offering on the floor near the picture before leaving."[14]

One dominant logic of the museum context has been education, embodied in the systematic ordering of the objects in a chronological and/or national sequence. We walk through the rooms of a museum, where the pictures are lined up on the walls in a determined sequence. "Viewing the pictures sequentially as we move from room to room, we follow the room numbers, the centuries, the schools. Insofar as the museum becomes pure path, abandons the dense spatial rooms of what were once *homes*, or, of course, the highly sophisticated space of a cathedral, it becomes a more perfect image of history."[15]

Even in the case of the Boston Museum of Fine Arts, which by the end of the nineteenth century had abandoned the goal of education for the more rarified pursuit of pure aesthetic contemplation, the historical arrangement by period and nationality remained in force.[16]

Barnes was only one of a number of rich collectors of this period – the later nineteenth and early twentieth centuries, when wealthy Americans began to use their riches to transfer much of the great art of the Old World to the New. It was fairly common for these collectors to construct homes and galleries in which to house their collections, often in the same space.

Isabella Stewart Gardner was among the first rich Americans to collect Renaissance art, often with the help of her agent, Bernard Berenson, and in 1903 she built a museum modeled on a Venetian palace, in the Fenway district of Boston, quite close to the Boston Museum of Fine Arts. She lived on the top floor of the museum and, at her death in 1924, her will created an endowment of $1 million for the support of the museum, including the stipulation that the permanent collection not be significantly altered. The museum has remained much as Mrs. Gardner left it, although in 2011 the museum added a new building adjacent to the original, which actually permitted the undoing of several modifications that had been made in the interim.[17]

Pittsburgh industrialist Henry Clay Frick built a mansion just off New York's Fifth Avenue on 70th Street that was designed to take into account his intention to leave his house and his art collection to the public. When Frick died in 1919 he left the house and all of the works of art in it together with the furnishings ("subject to occupancy by Mrs. Frick during her lifetime") to become a gallery called The Frick Collection: "I want this collection to be my monument."[18] He provided an endowment of $15,000,000 to be used for the maintenance of the Collection and for improvements and additions. After Mrs. Frick's death in 1931, the trustees commissioned John Russell Pope – who later designed the National Gallery of Art – to make additions to the original house. In December 1935 The Frick Collection opened to the public.[19]

The scion of another Pittsburgh fortune, transplanted this time to Washington, DC, Duncan Phillips turned part of his Dupont Circle mansion into a gallery for his growing collection of modern art in 1921. As the collection grew, and the public interest in viewing it, the family moved to a new home in 1930, converting the mansion into a museum. Over time, before Phillips' death in 1966 and since, additional galleries have been added to the museum, including 30,000 square feet in 2006.

There are others, of course, but these will suffice to note the pattern: rich collectors who transform their collections into museums, often literally built out of, or on top of, their palatial homes. Gardner, Frick, and Phillips were all fortunate that their homes-turned-museums were well located in stable, wealthy neighborhoods, and that the endowments they bequeathed were sufficient to maintain their collections more or less as they desired. Thus, in the best tradition of princely patronage, their memories are permanently entwined with the artistic value of the works they were able to purchase or, in some instances, commission.

Other collectors have not been so lucky, and one in particular should concern us here. John G. Johnson, like Barnes a self-made, wealthy Philadelphian, amassed a significant collection of early-Renaissance Italian primitives, Spanish, Flemish, and Dutch paintings that he exhibited in a house adjacent to his mansion at Broad and Lombard Streets in Philadelphia. At his death in 1917, Johnson left the home and the collection to the City of Philadelphia, on the condition that the collection be housed in the former residence and "maintained as a museum – a Public Museum – to stand pretty much as it will be at my decease."[20] Johnson further directed that "the residence shall be forever kept up and maintained as such Museum in which my art objects shall be exhibited."[21] If the home fell into disrepair, the will directed that the gallery was to be restored or rebuilt on the same site. However, when during the Depression the funds from the estate became insufficient to maintain the residence, the trustees closed the gallery and moved the art to the now-completed Philadelphia Museum of Art. The trustees then petitioned the orphan's court to permit the move to be made permanent. The court noted that the move was not only a flagrant breach of their trust, but also that the move was fait accompli. The court acknowledged the clear violation of Johnson's wishes, but weighed against this their judgment that "the neighborhood of Broad and Lombard Streets is not at the present time a desirable place to rebuild the Johnson property as a museum in which to exhibit the collection . . . [and] the Art Museum on the [Ben Franklin] Parkway is an excellent place for such exhibition."[22] So, while the court noted that the collection belonged to Johnson and "he had the exclusive right to dispose of it in any lawful manner he chose," they did not enforce the collector's wishes when they conflicted with other public interests.

Thus, in a strikingly similar case, also adjudicated in Pennsylvania, the court permitted the setting aside of the explicit wishes of the collector. In the view of legal scholar Seth Goldman, the Barnes case presented a less stringent case than the Johnson case. The Barnes Trust indenture contains the following clause:

> Should the said collection ever be destroyed, or should it for any other reason become impossible to administer the trust hereby created concerning said collection of pictures, then the property and funds contributed by [Barnes] to [the Barnes Foundation} shall be applied to an object as nearly within the scope herein indicated and laid down as shall be possible, such application to be in connection with an existing and organized institution then in being and functioning in Philadelphia, Pennsylvania, or its suburbs.[23]

The most legally compelling argument in favor of moving the collection was the financial situation of the Foundation, which was hampered by Barnes' own stipulations governing the handling of the endowment.[24] In his indenture, Barnes restricted the investment to federal, state, and municipal bonds. "Over time, this restriction severely eroded the endowment . . . For a collection housing what would today be billions of dollars in art, such an endowment would be slight compared to the need to safeguard the artwork against decay, theft, and litigation while still upholding the educational mission of the foundation."[25]

The Philadelphia Museum was not alone in desiring to acquire important collections in order to build its holdings, and many museums found themselves acceding to donors' wishes to have their collections serve as perpetual memorials maintained in segregated galleries. One Boston Museum curator referred to these self-contained collections housed in dedicated spaces as cemetery lots.[26] The most controversial instance of a major museum catering to the whim of an important collector is probably the Metropolitan Museum of Art's construction of a wing dedicated to the Robert Lehman collection.

Building on the art amassed by his father, Philip Lehman, his predecessor as chairman of the Lehman Brothers banking firm, Lehman owned one of the country's largest and most valuable collections of European old masters. The Metropolitan put him on the Board of Directors but, significantly, never made him president, apparently because he was Jewish. Relations between the Met and Lehman were often strained and in 1959 he removed works that had been lent to the museum and said that he would not leave them the collection after his death. Instead, because "the collection is unique and can stand on its own . . . [he thought] it would be best to keep the collection in the family mansion, on Fifty-fourth Street."[27]

Ultimately, Met director Thomas Hoving was able to ingratiate himself with Lehman – he arranged for the trustees to create the position of chairman of the board for him – and he persuaded Lehman to donate the collection, to be housed in a special wing built for it. The original proposal was to recreate the Lehman mansion as a wing "plunked down in the middle of the public parking lot on the south side of the museum."[28] In the end, the Lehman Wing was built as a new interior space that combined "period rooms" reproducing rooms in the mansion, with more familiar galleries. The Lehman Wing attracted a lot of criticism for the museum's willingness to cater to the collector's desire to memorialize his experience of living with the works of art, to treat the collector, in a sense, as an artist whose creations are the rooms in which he displayed his possessions. *New York Times* critic Hilton Kramer decried the museum's "surrender to a collector's fantasy, pride and will. It violates the whole spirit of modern museology, which aims to separate the art object from the accidents of ownership and let it stand permanently free in its own universe of discourse."[29]

In Hoving's account, on his deathbed Lehman acceded to Hoving's suggestion that the collection be kept separate for only 25 years, after which it could be dispersed, with his name attached, to the other galleries of the museum.[30] It's a nice story, but this hasn't happened. What has happened, though, is that when the original curator of the collection retired, his successor made some rearrangements in order to "provide rational groupings of the artworks from room to room so they can better be 'studied and experienced'."[31] For example, Sienese paintings of the fourteenth century have been hung together and "[n]ow they're no longer scattered through the rooms like decorative objects but treated like works of art."[32]

Once again, who speaks for the art that was bought by rich Americans taking advantage of the fallen fortunes of European aristocrats who, it should be noted, mostly inherited the works from ancestors who often acquired them through less than admirable means? Where does the interest of the artist come into play, if at all? Why should we privilege what Kramer calls the accident of ownership, or allow the taste of the collector to be celebrated as an aesthetic accomplishment and permanently enshrined in museum galleries? Certainly, the claims of public interest – and the museums in question all operate as tax-exempt public institutions – might better be served by allowing the presentation of art works to be determined by the best of contemporary art scholarship, acknowledging that future scholarship might lead to other arrangements and interpretations.

But this argument does not suffice as a response to the particular circumstance of Dr. Barnes and the Barnes Foundation. Barnes was not just another rich man accumulating treasures as a testament to his wealth and leaving a museum as a monument. He had other goals and other interests to pursue, and these deserve some attention.

When Barnes began collecting art in 1911, initially under the guidance of his boyhood friend, painter William Glackens and later Gertrude Stein's brother Leo, he focused his attention on contemporary painters rather than the old masters favored by most rich American art collectors. Barnes was quickly attracted by Renoir and Cezanne – by 1914 he owned 25 Renoirs and a dozen Cezannes, as well as 12 Picassos. By the time his collecting was halted by World War I, he owned works by Bonnard, Monet, Pissarro, and Sisley, as well as older works by El Greco, Goya, and Delacroix.[33] In the 1920s he resumed his shopping sprees, becoming a major force in the Parisian art market, known for driving a hard bargain. "I just robbed everybody," he once gloated. "Particularly during the Depression," he said, "my specialty was robbing the suckers who had invested all their money in flimsy securities and then had to sell their priceless paintings to keep a roof over their heads."[34] In the end, the Barnes Foundation owned 69 Cezannes, 60 Matisses, 44 Picassos, and, wait for it, 181 Renoirs. In addition, there are major works by Seurat, Van Gogh, Modigliani, Rousseau, Degas, Pascin, de Chirico, and Soutine. There are numerous works of medieval, Renaissance, and Baroque art, African sculpture, and decorative metalwork and furniture.

But, if the goal of wealthy Americans collecting European art was to burnish their social standing by association with masterpieces of the past, Barnes did not fit this mold. Rather, he was an evangelist of modernism and a champion of art, far from admired and valued by the guardians of respectability and social prominence. When he loaned 75 works for an exhibition at the Pennsylvania Academy of Fine Arts in 1923, the works were viciously condemned by the Philadelphia critical establishment as an infestation of an infectious scourge and trash unsuitable for the Academy. This was the beginning of Barnes' feud with the art and academic establishment of Philadelphia. At the same time, Barnes was building his Foundation in the suburb of Merion, and his estrangement from the art institutions of Philadelphia became an animating feature of his unique creation.

Barnes did not intend to build a museum for his collection, at least not in the mode of the museums then emerging in Boston, New York, Philadelphia, and other American cities. Barnes had a different model in mind, one that departed radically from the dominant linear, chronological ordering found in most art museums of his day, and of ours. Rather than a museum in the usual sense, Barnes saw his collection as an educational institution, as a school, with access mostly restricted to its students, rather than as a museum, which would have been open to the public.

This educational mission was not a new idea for Barnes, as he had earlier incorporated education into the operation of the A.C. Barnes Company, manufacturers of the important drug Argyrol. He decided that a six-hour work day would suffice for the operations of the company, and he mandated that the other two hours be devoted to a discussion group that he ran. This study group constituted an educational enterprise in which the workers read and discussed works by William James, Freud, and John Dewey.[35] Dewey, in particular, became a friend and mentor, and Dewey's ideas about art and education were both influenced by, and an influence on, Barnes' ideas and practices.

The Barnes Foundation that opened in 1925 was run as a school, with the classes mostly taught by Barnes himself. In his 1925 book, *The Art in Painting*, dedicated to John Dewey, he aimed to "set forth briefly the salient features of a systematic study of both old and modern paintings." The method proposed by Barnes "stipulates that an understanding and appreciation of paintings is an experience that can come only from con tact with the paintings themselves . . . It offers something basically objective to replace the sentimentalism, the antiquarianism, sheltered under the cloak of academic prestige, which make futile the present courses in art in universities and colleges generally."[36]

Visitors to the Barnes Foundation, in Merion or the new installation in downtown Philadelphia, used to seeing paintings lined up chronologically in a single array along museum walls, quickly note the idiosyncratic arrangement employed by Dr. Barnes. Here, in what he called wall ensembles, the paintings are hung in crowded arrangements, laid out by size in bilateral symmetry, often in two rows, and interspersed with numerous wrought-iron objects – hinges, door latches, etc. There is no familiar order of painters by period or country, but rather a series of arrangements that appears to follow different principles from room to room. Sometimes the point seems to be affinities of color, sometimes of shape, as when the contour of a Cezanne landscape of Mont Sainte-Victoire might echo the top of a portrait subject's bald head.

These arrangements are more than illustrations for Barnes' particular approach to teaching art to the Foundation's students, though they were also that. Over the years, especially after Barnes' death, they took on the character of independent aesthetic contributions.

As Jeremy Braddock notes: "A 1983 essay written by Barnes's protégé and collaborator Violette de Mazia on the subject of the arrangements is especially instructive in this respect."[37] He points to the way de Mazia argues Barnes' arrangements, which she names "wall pictures," should be seen as works of art in themselves:

> Each wall . . . , in the particular organization of its particular contents of paintings and other objects, makes up a particular "wall-picture" in the

> same sense that the artist-painter's particular organization of particular color units makes up his particular picture . . . Moreover, the paintings and other objects . . . are so hung that . . . they offer examples of certain compositional organizations found also in individual paintings and of other aesthetic features of concern to the artist-painter.[38]

In other words, Barnes came to be seen, by his disciples, if possibly not by himself, as an artist whose *magnum opus* was not merely the collection that embodied his extraordinarily good eye and exquisite taste, as might be said of Isabella Stewart Gardner, Henry Clay Frick, Robert Lehman, and other collectors of the era. No, Barnes can also be celebrated as the creator of these unique arrangements that themselves deserve the status of artistic works.

This, presumably, helps clarify what Julian Bond meant when he said: "[Barnes] will be violated in the experience he wanted you to have. And that's important, because it was *his* art. He had the right to do with it as he chose. And these people – these vandals – stepped in and took it away from him."[39]

But what, precisely, did the vandals take away from Barnes? Undoubtedly, they removed the collection from its home in the building he ordained in Merion: a building that most visitors agree was a unique and wonderful treasure, if also often relatively inaccessible and invariably idiosyncratic. But, as critics have almost universally agreed, the new museum has faithfully reproduced the layout and arrangement of the galleries, with Barnes' "wall pictures" recreated exactly as he laid them out, except with improved lighting and better climate control.[40]

So what has been lost in the move, or, more to the point, what is the trade-off between gain and loss? The old Barnes was a bit like a wonderful small restaurant that you treasure, and wish will thrive, but that you are also reluctant to tell others about because it won't be able to remain special if it becomes too popular. Once the secret of the Barnes was out – and this is another long and complicated story – the challenge of how to accommodate the growing number of visitors became intractable (and this is separate from other complicating factors, such as the need for substantial funds for upkeep of the building and preservation of the art).

It is also important to note that the Barnes collection contains a great many works of art that were not hung in the Foundation galleries and thus have been completely inaccessible to the public and to art historians and scholars. There is a good chance that we will learn of the existence of important works that have been obscured from view because they did not fit Dr. Barnes' designs for his "wall pictures" but that certainly merit our attention. The new building contains a temporary exhibit space where, it seems likely, the curators will be able to mount exhibits of these previously unseen works of art.[41]

Barnes was famous for refusing requests by art critics, academics, and various eminent folks who wished to see his collection, but at the same time readily inviting factory workers who asked to visit:

> As a young man, the novelist James Michener spent an enjoyable day with Dr. Barnes after he mailed a letter from Pittsburgh claiming to be a poorly educated worker in a steel mill. Michener had been denied entrance as a Swarthmore student on three previous occasions. After he became famous, Michener bragged about his coup—and became, in his words, "a target of Barnes's scatological abuse for a five-year period."[42]

The Foundation indenture instructs the trustees to ensure "that the plain people, that is, men and women who gain their livelihood by daily toil in shops, factories, schools, stores and similar places, shall have free access to the art gallery and the arboretum upon those days when the gallery and the arboretum are to be open to the public." The same clause goes on to specify: "On Sunday of each week during the entire year the gallery and the arboretum shall be closed to students and public alike."[43]

This contradiction between the ostensible invitation to plain people and the barrier imposed, not only by its somewhat remote and obscure location, but also by the lack of access on the one day most of the plain people might be able to visit, was not unfamiliar in the history of American museums. "For the first twenty years of its existence, [the Metropolitan Museum in New York] was closed on Sundays, despite massive political and popular demands that it be open on this, the only day that most working people could go."[44]

Despite Barnes' apparently sincere wishes for his Foundation to transform the artistic education of the plain people, over the decades its classes – offered only on Tuesday, during working hours – became the province of educated, middle-class folks, many of them Main Line housewives and retirees.

After the death of Dr. Barnes in a 1951 traffic accident, the Commonwealth of Pennsylvania, urged on by Walter Annenberg's *Philadelphia Inquirer*, waged an ultimately successful campaign to extend the times that the museum would be open to the public. In 1960 the State Supreme Court ordered that the Foundation be open to the public, by appointment, on Friday and Saturday of each week. While this certainly increased the possibility of access to the collection, it hardly changed the basic remoteness of the art from the plain people. I can readily recall visiting the Foundation many times in the 1970s and 1980s, usually without a reservation, and generally finding the galleries pleasantly uncrowded.

The story of the Barnes Foundation's internal and external struggle in the Glanton era and after is a long and complex tale that others have told,[45] and it's not my present focus. What is relevant here is that, after the extensive publicity engendered by the worldwide tour of a large number of paintings from the collection, the demand to visit the Foundation brought on an understandable and intractable conflict with the neighbors in what had been a quiet, upscale Main Line community. At the same time, the powerful forces of Philadelphia philanthropy, the Pew Trust and the Annenberg and Lenfest

Foundations, saw an opportunity to add a world-class art collection to the grouping of art institutions in central Philadelphia.

In the conflict between the remaining but fervent disciples of the Barnes Foundation tradition and the combination of political, economic, and cultural power of the establishment, the outcome was probably never in doubt. Certainly a unique and special place no longer exists, but it is not as certain that Dr. Barnes' true mission has been subverted. To the extent that his real goal was to bring art – in his own special arrangement and manner – to "the plain people" rather than merely to please the critical and academic establishments, then perhaps his goal is being realized despite the efforts of his defenders. After all, even though it includes a cafeteria and a shop that will probably sell tchotchkes as well as Barnes' own books, even though it will have spaces that will permit the sort of social gatherings and fundraisers Barnes explicitly forbade, the galleries that faithfully reproduce his "wall paintings" will be open to the public six days a week, and Friday evenings, in the center of Philadelphia. Certainly, it will be much more accessible to the public, including "men and women who gain their livelihood by daily toil in shops, factories, schools, stores and similar places."

It is likely the case that there was an irresolvable conflict between Barnes' mutually contradictory impulses to reach the public and to restrict and control access to the amazing collection he amassed. But, if you factor into the equation the interests of the artists whose work he purchased, who, it might be imagined, would have liked their work to be seen by large numbers, then the balance might tip in favor of access. And then there is the truly unanswerable question, one that can't be resolved by equating the collector with the paintings: who speaks for the art?

Notes

1 Don Argott, *The Art of the Steal*, MPI Home Video (2009).

2 The Barnes Foundation, created in 1922 by Dr. Albert Barnes, contains one of the world's greatest collections of post-impressionist art, as well as other masterpieces and valuable craft works. The collection has been valued at many billions of dollars.

3 Quoted in James Panero, "Outsmarting Albert Barnes," *Philanthropy Magazine* (Summer 2011), www.philanthropyroundtable.org/topic/donor_intent/outsmarting_albert_barnes.

4 Henry Hansmann and Marina Santilli, "Authors; and Artists' Moral Rights: A Comparative Legal and Economic Analysis," *The Journal of Legal Studies* 26 (January 1997): 95.

5 "Take the two paintings which are still the most expensive ever sold: Van Gogh's *Portrait of Dr Gachet*, and the equally famous Renoir *Au Moulin de la Galette*. These were bought together, at the peak of the boom, for $240 million, a price that staggered even the boldest art speculators.

The purchaser was one Ryoei Saito, 77, then the billionaire chairman of the controversial Daishowa paper manufacturing company. Mr. Saito promptly outraged the art world by declaring that when he died he would have the paintings cremated along with his body – so that his heirs would not have to pay billions of yen in death duties.

As it turned out, he never got the chance. With his company teetering on the brink of bankruptcy, Mr. Saito was deposed and when last seen was being driven into a police station with his head buried in his arms to be charged with bribery and corruption. The paintings have never been seen publicly since, and rumour has it they are also in a bank vault.

A survey by the Nikkei news organisation of just five of Japan's scores of non-bank financial institutions found that they were sitting on a total of $1.5 billion-worth of confiscated art. The real total is anyone's guess – the banks certainly aren't owning up Ben Hills, "Raiders of the Lost Art: Japan's Great Art Bust. Why a Lot of the World's Great Art Has Disappeared," *Sydney Morning Herald*, March 12, 1994, http://benhills.com/articles/japan-unlimited/raiders-of-the-lost-art/.

6 *Commonwealth vs. Barnes Foundation* (1960): 502–503.

7 Barbara Rose, "Calder's Pittsburgh: A Violated and Immobile Mobile," *ARTnews* (January 1978), 39.

8 John Merryman and Albert Elsen, *Law, Ethics and the Visual Arts* (Philadelphia: University of Pennsylvania Press, 1987), Vol. 1, 160.

9 Rosalind Krauss, "Changing the Work of David Smith," *Art in America*, 30 (1975): 31.

10 Hilton Kramer, "Questions Raised by Art Alterations," *New York Times*, September 4, 1974.

11 When he died in 1968 Brod left the original manuscripts, many still unpublished, to his secretary (and, probably, his mistress), Esther Hoffe. She sold some of the documents, and after her death in 2007, the remaining documents were inherited by her daughters. There ensued a legal battle over the papers, which are kept in Swiss and Israeli bank vaults as well as in the Tel Aviv apartment her daughter Eva shares with an unknown number of cats, between the daughters and the Israel National Library, which claims ownership in the name of the Jewish people and all humanity. See Elif Batuman,"Kafka's Last Trial," *New York Times Magazine*, September 22, 2010, www.nytimes.com/2010/09/26/magazine/26kafka-t.html?pagewanted=all. In late 2012 an Israeli judge ruled that the documents must be turned over to Israel's National Library, which plans to publish them online. Jodi Rudoren and Myra Noveck, "Woman Must Relinquish Kafka Papers, Judge Says," *The New York Times*, October 15 2012, A9.

12 Philip Fisher, "The Future's Past," *New Literary History* 6 (1975): 587–606.

13 Fisher, "The Future's Past," 590.

14 Fisher, "The Future's Past," 593.

15 Fisher, "The Future's Past," 591.

16 Paul DiMaggio, "Cultural Entrepreneurship in Nineteenth-century Boston, Part II: the Classification and Framing of American art," *Media, Culture and Society* 4 (1982): 303–332.

17 Nicolai Ouroussoff, "An Architect Pays Respects to a Dowager," *The New York Times*, January 20, 2010, www.nytimes.com/2010/01/21/arts/design/21gardner.html?pagewanted=all.

18 Carol Duncan, *Civilizing Rituals: Inside Public Art Museums* (New York: Routledge, 1995) 75.
19 www.frick.org/collection/history.htm.
20 Quoted in Seth Goldman, "Just What the Doctor Ordered? The Doctrine of Deviation, the Case of Doctor Barnes' Trust and the Future Location of the Barnes Foundation," *Real Property, Probate and Trust Journal* 39, no. 4 (2005): 738.
21 Goldman, "Just What the Doctor Ordered," 738.
22 Goldman, "Just What the Doctor Ordered," 739. The court also noted "the tremendous change that was taking place in the physical appearance of central Philadelphia," and the fact that "the city was anxious to build a health center on the [Broad Street] site."
23 Goldman, "Just What the Doctor Ordered," 727.
24 The Foundation's finances were also challenged by the actions of Richard Glanton, when he was president of the Foundation, in suing the town of Merion and spending enormous amounts in legal fees. John Anderson, *Art Held Hostage: The Battle over the Barnes Collection* (New York: Norton, 2003).
25 Panero, "Outsmarting Albert Barnes."
26 Cited by Duncan, *Civilizing Rituals*, 60.
27 Jane Boutwell, "The Talk of the Town," *The New Yorker*, June 9, 1975, 31.
28 Thomas Hoving, *Making the Mummies Dance: Inside the Metropolitan Museum of Art* (New York: Simon & Schuster, 1993), 157.
29 Quoted in Duncan, *Civilizing Rituals*, 69.
30 Hoving, *Making the Mummies Dance*, 162.
31 Grace Glueck, "New Look for an Enclave of Art," *The New York Times*, April 11, 1989.
32 Glueck, "New Look for an Enclave of Art."
33 Howard Greenfield, *The Devil and Dr. Barnes: Portrait of an American Art Collector* (New York: Penguin Books, 1987), 46.
34 Quoted in Panero, "Outsmarting Albert Barnes."
35 Anderson, *Art Held Hostage*, 23–24.
36 Albert Barnes, *The Art in Painting* (Merion, Pa.: The Barnes Foundation Press, 1925).
37 Jeremy Braddock, *Collecting as Modernist Practice* (Hopkins Studies in Modernism), (Baltimore: Johns Hopkins University Press, 2011), 103.
38 Braddock, *Collecting as Modernist Practice*, 105.
39 Argott, *Art of the Steal.*
40 The one exception to the exact replication of the Foundation's layout is the placement of Matisse's *Joy of Life*, one of the most important works in the collection. As Braddock put it, Barnes had the painting "hung inconveniently at the top of the stairwell leading to the second-floor galleries. Though it is reasonably well lit, close inspection of the painting is very difficult," Braddock, *Collecting as Modernist Practice*, 111. In the new building the painting is set off in a separate room where it can be easily seen.
41 Keep in mind that Dr. Barnes forbid the art works he owned from being reproduced or loaned, although his own books did contain (fairly poor) black and white images, and there is as yet no available complete color catalog of the collection.

42 Dilys Winegrad, "A Survivor's Saga: Growing up and Moving on with the Barnes Collection," *Broad Street Review*, October 1, 2012.
43 The Barnes Foundation's Bylaws, Clause of April 30, 1946, www.barneswatch.org/main_bylaws.html.
44 Duncan, *Civilizing Rituals*, 56. And, when the Met in 1891 finally opened on Sundays, *The New York Times* noted, with relief, that it did not, as had been feared attract, "Essex Street Polish Jews and Thirty-ninth Street and Eleventh Avenue [Irish] hood carriers in ragged clothing and dilapidated hats," Duncan, *Civilizing Rituals*, 59.
45 See Anderson, *Art Held Hostage*, for an account.

Part II

Memorializing images

Chapter 6

Introduction

Residual/visual: images and their specters

Kevin Gotkin

When spirit photographers of the late nineteenth century led their clients to believe they were capturing ghosts during portrait sittings, they were in fact manipulating residues. They laid various light-sensitive chemical substances such as iodized silver and silver chloride on top of a metal plate before exposure and fixation of their images. Capitalizing on the public's mystification about the finer details of the newfangled medium, spirit photographers often used plates that had already been exposed to a different scene, yielding an eerie specter in what was actually a doubly exposed image.[1] Where once liquids were essential for the production of images, they've evaporated from the way we make images now. Getting them too close to cameras, in fact, often spells technical disaster, as anyone whose camera has been soaked by the rain will know.

And yet the residues that were once literal for the creation and manipulation of images are a nice figurative frame for inquiries into how we memorialize images and how images memorialize our social world. Memory is hard to hold, slippery when we try to contain it, yet an ever-present force in our visual culture. Images seem to leave residues in their wake that in turn awaken experiences for viewers where they circulate. Spirit photography, for example, embeds memories of the rise of spiritualism in the late nineteenth century and its rather sticky confrontation with men of science bent on exposing their slippery theologies as fraudulent. Often the residues images create are wholly intangible, exist in levels of interaction not easily apprehended in traditional language vectors, and therefore heighten the sense that images are leaky. When images seem to activate things that are beyond the sum of their parts, it is often their residues we sense. Memory, memorialization, and history consist in these image residues and thus it is these residues to which we must orient ourselves.

Perhaps more directly related to the memory work of images are not the liquids used in the production process but the specters purportedly captured by them. What Barthes called the *punctum* of an image[2] – its punch in the gut of someone looking at a photograph of a dead relative – is the residual ghost created by the gap between the materiality indexed by an image and the cultural orbits that circle around it. This need not be confined to the experience of an individual. Iconic photographs of war serve to remind us of the ghosts whose stories were enfolded into world historical developments that far exceeded particular individuals. The meanings embedded into widely circulated images create collective residues, liquid as they are, bleeding into other pools of thought. Like image residues, specters float in and out of our analytic frames.

The fluidity of memorialized and memorializing images is what makes it so hard to neatly capture how images, history, and memory tessellate. The containers we use to try to enclose particular historical spills often need to be washed out to make them continually useful to the task of explaining how images memorialize. The following chapters try out new configurations for what these containers might look like, alternating between business history, unintentional image overflows, and ambiguous image reference-markers.

Marita Sturken's chapter on the success of Facebook after the demise of Kodak and Polaroid is a careful multi-faceted history that shows us how businesses have shaped emotion (nostalgia and memory chief among them) in the design and marketing of the products we use to document our personal lives. It is no accident, Sturken shows us, that Facebook's Timeline (a function that puts a user's personal information into a scrapbook-like chronological history) debuted not long after Kodak and Polaroid went out of business. Facebook's attempts to manage how users create meaning around their family memories explicitly harken back to the work that Kodak and Polaroid did in shaping the cultural meanings around home photography. This business model, Sturken tells us, is a reincarnation of the "razor-blade strategy" used by Polaroid and Kodak, so no wonder Facebook's marketing for Timeline makes a direct historical allusion to a classic Kodak ad. In reading the histories of these companies' business models with cultural histories of home/amateur photography, Sturken shows us how the technologies we use to document our lives are diverging from the meanings handed to us from twentieth-century photographic practices just as they are steeped in them.

Ira Wagman's chapter picks up, in a sense, where Sturken leaves off. He deals with some of the ramifications of the shifts that Sturken is describing, focusing in particular on moments of rupture, when images induce feelings of embarrassment, guilt, and shame. These are not trivial or merely peripheral considerations for the way images traffic in contemporary settings. He shows how the anxieties and real consequences that emerge around online embarrassment plunge us into questions about global rights (to be or not to be forgotten). In touring the various remedies proposed for dealing with the aftermath of what we might call "image trouble," he finds that none come without deep tradeoffs. He turns to consider possibilities born *out of* images' affective resonances and not *despite* them. Forgiveness, he suggests, might be an ethical scaffolding for a digital future that needs help being constructed.

In the final chapter of this part, Roderick Coover adds an important keyword to the way we understand the indexical quality of images: ambiguity. What images purport to document is often less sure than we imagine, as the conditions of authority and evidentiary capacity give us, as Coover says, "[a]rt and science blurred." In his reading of a number of images, Coover shows us the unsettled status of what we call the "documentary" image. That is, the "document" in "documentary" is not always a straightforward one-to-one correlation between a worldly phenomenon and its mimetic capture in media. At the close of his chapter, Coover suggests that the digital turn is encouraging a wider recognition of images' fundamental ambiguity, which asks us to reckon with new kinds of relationships between producers and users of images.

In particular, the ghosts in this section deal with the affective weight of images. These chapters deal with what images – and their ethical and technological circumscriptions – *feel* like as they course through networks, across platforms, and into the fabric of social life in our digital age. Affect, understood here as the complex relationship between mental interiorities of desire and external structures of feeling, is an important contributory to the way images memorialize. When images accrue meaning through their production and dissemination in digital venues, they create relations of care between many actors, beyond, let's say, the sender and receiver of a photo on a cell phone. As the following chapters show us, these webs account for the affective weight we often sense in dealing with images in even quotidian ways.

Though none of these chapters deals explicitly with ghosts, we can see in them what Derrida once called a "hauntology": in the mire of postmodernism, we orient ourselves more toward a flickering "ghost" of the past than a certain sense of what that past consisted of. Presence becomes a liminal existential category, where analytical structures must deal with the duplicities and contradictions of postmodernist thinking. Thus, Sturken, Wagman, and Coover render some of the specters that hang over our attempt to reckon with the wily nature of contemporary images.

Memory is one of many things we do with images, but it is an important one. Just as residues and specters float and leak, so does memory wash over and through our contemporary visual culture. The memories embedded into images remind us that contemporary image-making is never utterly contemporary; there are historical transferences and (dis)continuities that relay between production and dissemination of images. This part pushes us to find interesting ways to account for these productive slippages and, in so doing, model a capacious framework for understanding the intersection of memory, history, and visuality.

Notes

1 For a comprehensive treatment of spirit photography and its historical context (specifically in Britain), see Jennifer Tucker, *Nature Exposed: Photography as Eyewitness in Victorian Science* (Baltimore: Johns Hopkins University Press, 2005).

2 Roland Barthes, *Camera Lucida* (New York: Hill and Wang, 1981).

Chapter 7

Facebook photography and the demise of Kodak and Polaroid

Marita Sturken

In December 2011, Facebook introduced Timeline, a restructuring of Facebook user pages that reformatted their content – personal posts, photos, and videos – into a chronological framework. In the process, Facebook users were forced to have their pages shaped into personal histories, organized, like photo albums, yearbooks, and diaries, from birth to the present (the switch was voluntary for a period of time, and then mandatory). In the process, Facebook "conveniently" filled in the personal information that it had on users from their personal profiles, such as date of birth, hometown, schools attended, etc., producing a fair amount of discomfort in demonstrating the extent of their "knowledge" of users for whom privacy had become an increasing concern. Facebook has sold Timeline to its users as a kind of multimedia self-narration of one's "life story." Importantly, Timeline rearranges all previous Facebook activities into a format that is particularly photo-friendly, thus creating a context for personal histories to be narrated through photographs. The rolling out of Timeline was thus greeted in the technology press as Facebook claiming the terrain of photographic memory and the family album. "Like it or not, Facebook is the new Kodak," writes K. Kelleher. "For many Facebook users, Facebook is becoming the photo album of the 21st century."[1]

Facebook needed to sell the concept of Timeline to its users, and so it produced an ad to convince users that Timeline would facilitate the creation of personal photographic histories. In the ad, Andy Sparks, apparent Facebook employee, is born in 1974, grows up, graduates high school, meets a girl ("in a relationship" in Facebook parlance), gets married, and then has a little girl of his own. Along the way certain themes are established: Andy singing badly at a microphone; Andy as a cut-up; Andy as a goofy yet loving father. The ad races along Andy Sparks' life story as if we are flying over it, winging into the present (Figure 7.1).

The Facebook Timeline ad directly recalls one of the most famous television ads ever made, Kodak's "Turn Around" ad. The 1963 ad, set to the accompaniment of a Harry Belafonte–Malvina Reynolds song (sung by Ed Ames), is now considered to be a classic in the history of television advertising. It traces the childhood of a young girl, through an imagined photo

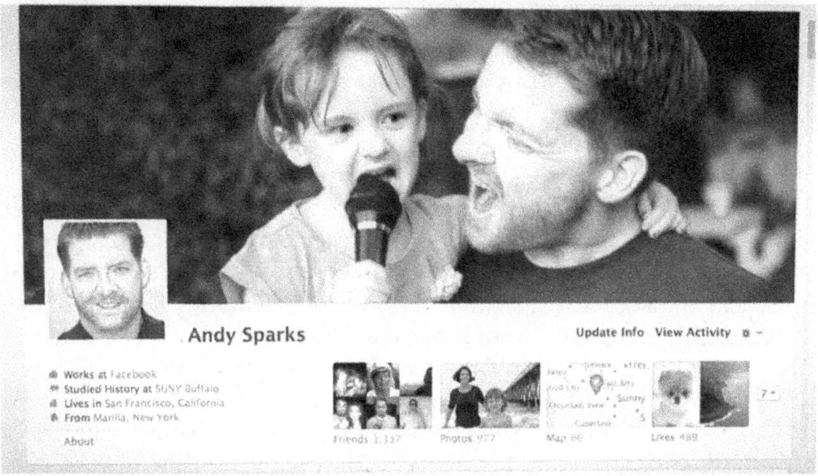

Figure 7.1 Still from the Facebook Timeline ad, 2011.

Source: www.youtube.com/watch?v=mR8x_xp4zhU.

album (the name Judy embossed on its cover), as she grows up and then, exiting childhood, "has babes of her own." At the end of the ad, the viewer is told that, "you can do it too, all it takes is a camera, Kodak film, and thoughtfulness." (It's worth noting what is really being sold here product-wise is apparently not a Kodak camera – just "a camera" – but Kodak film.) Like Andy Sparks, Judy is a real person, in this case Judy Ellis Glickman, whose father, Irving Ellis, took the photographs.[2]

"Turn Around" gained its place as a classic of television advertising because of the ad's capacity to depict emotional and familial bonds and then to establish the crucial role of the Kodak brand in maintaining those bonds. The ad thus succeeds in affirming that Kodak is the key conduit through which memories and familial relations are maintained, and it tugs at the viewer's emotional registers in order to align Kodak with the affect of family memories. This situates it within a longer history in which many communication technologies have aimed to sell commodities as the means through which individual relationships are reified, exemplified by such slogans as AT&T's well-known "Reach Out and Touch Someone."

The aim of a massive corporate entity such as Facebook (with over one billion users)[3] to become the photo album of the digital generation coincided with the demise of the two giants of the photographic industries of the twentieth century: Kodak and Polaroid. Polaroid declared bankruptcy several times between 2001 and 2009, and, in true ironic fashion, Kodak's declaration of bankruptcy in January 2012 was a mere month after the introduction of Facebook Timeline, and a few months before, in April 2012, Facebook paid $1 billion for Instagram. Instagram's iconography references both Kodak (the Instamatic) and Polaroid (the Instagram logo plays off the Polaroid camera and its most popular filter style – "1977" – looks like an SX-70). Around the same time, Google launched a campaign for Google Chrome that marketed it as an ideal format for family memories. In 2012, the technology press was thus replete with articles about how social media has become the new photo album. Yet, as a social-media platform, one that retains ownership over the content that is posted on its site and that profits from access to consumers and their personal data, Facebook's aim to position itself as the keeper of personal and familiar memories raises complex ethical issues about the intersection of capital and affect.

My aim in this chapter is to trace the intersection of these business histories (Kodak, Polaroid, Facebook), the family and personal photograph, and the selling of a particular set of personal and family photographic practices in order to consider the ethical issues raised by the relationship of corporate marketing to the very affective practices of family and personal photography. I thus aim to examine the history by which Kodak quite effectively shapes a set of technology practices of family photography throughout the twentieth century, which is then followed by the Polaroid Corporation with a different set of aims (the photo as instant and as entertainment) yet with equal market dominance and influence, and, with the demise of these two companies, Facebook then aims to shape personal photographic histories. It is at the juncture of sentiment, marketing, business strategy, and technological history that I would like to pose a set of questions about the ethics of corporate ownership and shaping of personal and family photographs.

One of the key business principles that shapes the history of Kodak and Polaroid, which has particular implications for the shift from analog to digital in the photographic industries, is the business model of the "razor-blade strategy." The razor-blade strategy refers to the ways that corporations sell consumers devices at low cost (or give them for free) that then allow these companies to sell over-priced products that are needed for continued use of those original devices. So, a razor is relatively cheap, but the replacement razor blades are expensive and constitute a substantial part of the company's profits in ways that are relatively hidden from consumers. The "razor-blade" business strategy can be seen in evidence today as the profit basis for contemporary products such as digital printers (cheap or often free with overpriced ink cartridges), cell phones (low-cost phones with high-cost calling plans), or console-based video games (inexpensive consoles, expensive games).

Kodak and Polaroid were some of the primary beneficiaries for many decades of this business strategy. They sold cheap cameras to consumers and then made the majority of their profits on film and (for Kodak) film processing. The profitability of this was quite extraordinary throughout most of the twentieth century. But when the digital emerged, effectively eliminating the cost of film and making it "free," this model could no longer hold, and neither company could adjust fast enough – as if each were a larger ocean liner unable to turn quickly to avoid the iceberg. While Kodak's effective demise has been more spectacular, given that its brand value was so deep, Polaroid suffered the same inability to be a flexible-enough company to survive the demise of the photographic-film business. In a 2008 talk at Yale School of Management, Polaroid's former CEO Gary T. DiCamillo famously said: "We knew we needed to change the fan belt, but we couldn't stop the engine. And the reason we couldn't stop the engine was that instant film was the core of the financial model of this company. It drove all the economics."[4]

The razor-blade strategy may seem quaint, but it has proven to have a certain kind of longevity in the digital age – one could chart this from the very early revelation by nascent e-mail providers such as Hotmail (in the mid-1990s) that they could give users free e-mail accounts in order to tap into their personal networks. Today, the "free" services of social media such as Facebook or platforms such as Google with Google Chrome and gmail, take the razor blade further, redefining it not as the value of the "secondary" product that consumers are sold to keep their devices functioning (such as film) but as the value of delivering users and their data to marketers and corporations. The service is "free" with the price of loss of privacy and the agreement to be under the gaze of marketers, constantly available to consumer messages. So, one could say that the cheap razor, the Instamatic, or SX-70 camera, and the Facebook and Google platforms are all product *portals*, intended to deliver consumers into an ongoing relationship with the brand and to create a bond that will then root consumers in a sentimental and affective relationship to a product or brand. The difference with social media is that the shift is from the value of the film needed to make the camera function to the value of the integration of brand messages into users' social-media lives. We thus go from razor blades to consumer data. With the increased role of photographs in social media, and the advent of platforms such as Facebook Timeline that intend to shape consumers' personal photographic histories into online photo albums, personal images are thus increasingly owned by corporations in the digital cloud.

In order to understand the implications of the terrain of social media's emergence as the key platform for personal and family images, it is necessary to make sense of how Kodak and Polaroid established a particular set of photographic practices. Thus, before turning to examine photographs in Facebook and Instagram, I would like to look more closely at the history of the success and demise of Kodak and Polaroid, and the different models of photographic practices that each company marketed and sold.

Kodak

The name Kodak itself has come to signify so many aspects of a particular history of photography: a modern name, created for the brand, a company founder and patriarch (George Eastman), a company town (Rochester, New

York), a brand so dominant that it defined an entire set of practices, and a set of familiar slogans that were so influential that they became catchwords of American culture: the "Kodak moment," "To Kodak," "You Press the Button, We Do the Rest," and so on.

The origins of the mass market of photography can be found in 1888, with George Eastman's patenting of the name Kodak and developing the Kodak No.1 camera. Kodak's history forms a parallel history to that of amateur photography, integral to it, with an arc of about 123 years until Kodak's declaration of bankruptcy in January 2012 and the dramatic transformation of amateur photography over the last decade in particular. As late as 1976, Kodak commanded 90 percent of film sales and 85 percent of camera sales in the US.[5] It has had a faint-hearted corporate afterlife, largely based on its valuable holdings of patents and intellectual property, but its effective demise is irrevocable.[6] There are business stories to be told here (the last few decades of Kodak's business profile has been told as a cautionary story about bureaucratic obstacles, misguided management practices, complacent business strategies, brutal downsizing, and lack of innovation or foresight aided by outmoded corporate structure),[7] stories of business philosophy (the razor-blade model based on the belief that film would continue to be a crucial consumer product),[8] personality stories (Kodak founder George Eastman, an eccentric inventor and a key patent holder, committed suicide in 1932 in a strangely meticulous fashion after accessing his potential future as an invalid and shooting himself in the heart),[9] and stories of technology (the development of each new Kodak camera from the Brownie to the Instamatic can be seen as a defining feature of the structures of feeling of each era). Within this history, Kodak played a significant role in establishing the practices of amateur photography as defined by family photography in particular.

Kodak's early campaigns introducing the concept of photography by the masses spoke to late nineteenth- and early twentieth-century consumers within the emergent values of modernity – as mobile citizens who had leisure time. With leisure as a newly understood category for modern workers, and with the growing population of young women moving through and working in urban centers, Kodak sold the idea of photography as freedom, thrills, and access to the great outdoors through its Kodak Gilmore Girls, who populated its early campaigns. In her insightful book, *Kodak and the Lens of Nostalgia*, Nancy Martha West makes clear that the idea of "Kodakery," the notion that the Kodak photograph was a tool of self- and family documentation, took longer to take hold.[10] She argues that Kodak succeeded, through its market monopoly and its hold on the public imagination, in helping to shape not only practices of amateur photography but also *actual practices* of remembrance and nostalgia themselves. Thus, Kodak begins with slogans that embrace leisure ("Kodak as You Go," "Take a Kodak With You," "All Outdoors Invites Your Kodak") and then, as it began to understand the role of women consumers to its product, it moves to slogans about family unity

("Let the Children Kodak" (Figure 7.2), "The Brownie Family," "Let Kodak Keep the Story").

Kodak also provided a radical new sense of image abundance. In the late 1880s, Kodak began manufacturing film with 100 exposures, which, as West writes, "was probably over ten times as many photographs as the average middle-class American family owned at the time."[11] This simple technological development transformed the practices of amateur photography by tapping into the "dominant hope of American culture since the early nineteenth century: effortless abundance."[12]

Kodak had such a dominance in the market that its influence on the emerging set of practices of consumer photography was considerable.

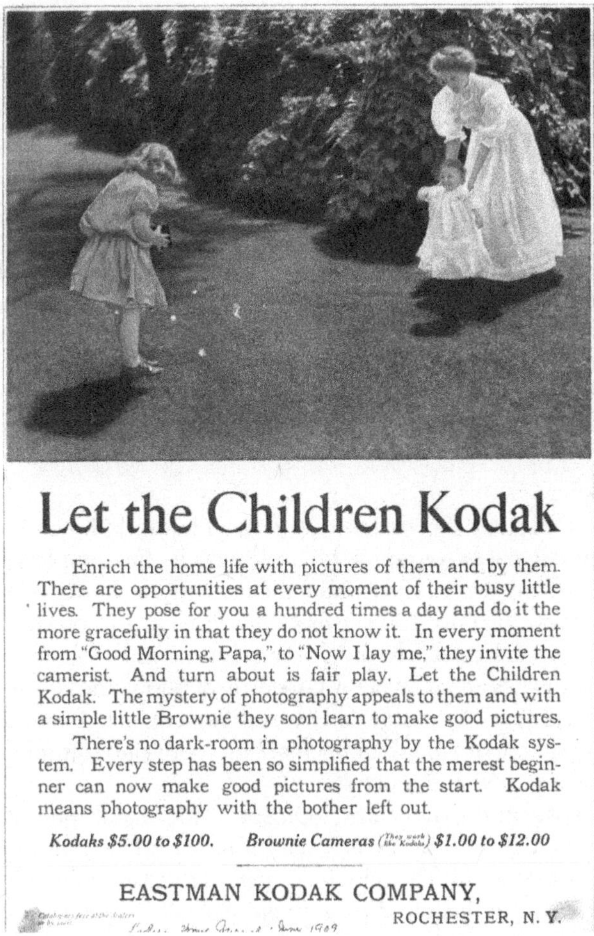

Figure 7.2 "Let the Children Kodak" ad, 1909.

Kodak was a key factor in teaching consumers the practices of amateur photography. In his analysis of snapshot photographic culture, Richard Chalfen refers to the "home mode of pictorial communication" as "Kodak Culture," acknowledging the dominance of Kodak in shaping the practices of the snapshot.[13] By the early twentieth century, Kodak had focused its attention on encouraging the production of abundant photographs of families and childhoods, all within a glow of nostalgia. Thus, Kodak helped to define the key role of amateur photography as domestic photography with its concept of "Kodak moments" – birthdays, childhood gatherings, family gatherings, and rituals that could form the *platform* for nostalgia. The culture of photography initiated by Kodak was thus not only about the capacity of the camera to produce an "abundance" if not an overproduction of images, but also about the broader implication of this concept of the "Kodak moment" – the structuring of life to create photographable moments and, by extension, the definition of difficult moments as the unphotographed, the forgettable. As West notes, Kodakery produced a narrowed view of what the snapshot photograph should do, instilling the notion that amateur photography was about the nuclear and private family: "it ensured that the nuclear family and the events that help maintain its survival – births, marriages, vacations, holidays – would form the dead center of photographic culture and, indeed, of culture as a whole."[14] Not only did Kodak establish this as the realm of photography, but also it encouraged and ultimately influenced consumers to effectively structure their lives with "Kodak moments" that were orchestrated and arrested in time, preserved in anticipation of nostalgic reflection. Hence, the genre of the photo album emerges with a collective cultural sense of the kinds of family pictures that should be taken and preserved within it.

Polaroid

If Kodak established the role of domestic photography in the consumer market of amateur photography, Polaroid was its hip counterpart. Polaroid began its company arc in 1937 and ended it with declaring bankruptcy several times from 2001 to 2009, eventually ceasing the production of film. Importantly, while Kodak's brand dominance and meaning about family pictures and Kodak moments were retained throughout the twentieth century, Polaroid was defined by instantaneity and later by art. This was not about the domestic sphere of Kodak moments, but about photography and parties, swinging culture, sex, and hipsters.

The concept that drove the development of the instant photographic camera, first invented by Polaroid's legendary founder Edwin (Din) Land in 1948, was that of "one-step."[15] The initial idea came from the combination of developing chemicals and celluloid in a single package. A camera roller would squeeze the chemicals onto the exposed film as it was processed out of the camera and, after a short moment, the consumer would peel off the chemical sheet in order to reveal the final image. The awkwardness of the Polaroid model – messy chemicals, the necessary and complex components of the camera to both take and process the picture, the excess garbage produced – points to the radical nature of its invention. This is intensely engineered instantaneity (*despite* rather than because of the technology), not the instantaneous ease of the digital image. Yet, in its values of instant (and sharing, as I will discuss) the Polaroid camera is prescient about the values of the digital image. Initially, the instant cameras were marketed as a means to improve image-taking – "Improve your pictures on the spot!" – but the values of entertainment and thrill soon emerged as Polaroid's marketing concepts – "there's no thrill like seeing your pictures 60 seconds after you shoot them."

The Polaroid SX-70, which hit the consumer market with immediate success in 1972–1973, was the product that most personified the Polaroid ethos and effect. The signature white borders of the SX-70 frames the instant photograph with a certain aesthetic. The SX-70 image develops over several minutes before the viewer's gaze (rather than under foil in the original film), thus creating an experience of suspense as it is gradually revealed. The SX-70 thus transformed the process of development (normally an inconvenience in which consumers – Kodak consumers – dropped film off and then waited days or weeks for its return) into something fun, potentially collective, if not titillating. "Take and show party pictures while the fun is going on," early Polaroid ads proclaimed. The connection of the Polaroid image to parties, celebrations, youthful gatherings, and collective viewing was thus both an intended and unintended consequence of its technological design and key to its cultural meaning as hip in the 1960s and 1970s.

Nowhere was Polaroid's aim for hipness clearer than in its campaign for the aptly named Polaroid Swinger in the 1960s (launched in 1965), which featured the camera as a kind of mod accessory (with a white plastic model and a viewfinder that lit up with "yes" when there was enough light for the picture). Here, the camera, as the portal to the expensive film, was sold for both its cheap price and its mod style – the lyrics of the ad jingle promised, "It's more than a camera, it's almost alive! It's only nineteen dollars and ninety-five." The ad for the Swinger featured (as did many ads in that era) young people frolicking on a beach and taking pictures of each other (Figure 7.3). The instant photograph was thus marketed as a youth product, one that was more about the social interactions and activity the camera produced than the pictures it took (the Swinger's photos were small and not particularly good).[16]

Edwin Land was a dominant inventor-CEO who had an obsession for product perfection at the level of engineering and a flare for product launch in ways that made him the prototype for Steve Jobs of Apple. Land introduced the SX-70 at a shareholders' meeting in April 1972 with a ten-minute film created by Charles and Ray Eames. Interestingly, at one point the film states that the "hope" of the SX-70 was to "change the person who takes pictures

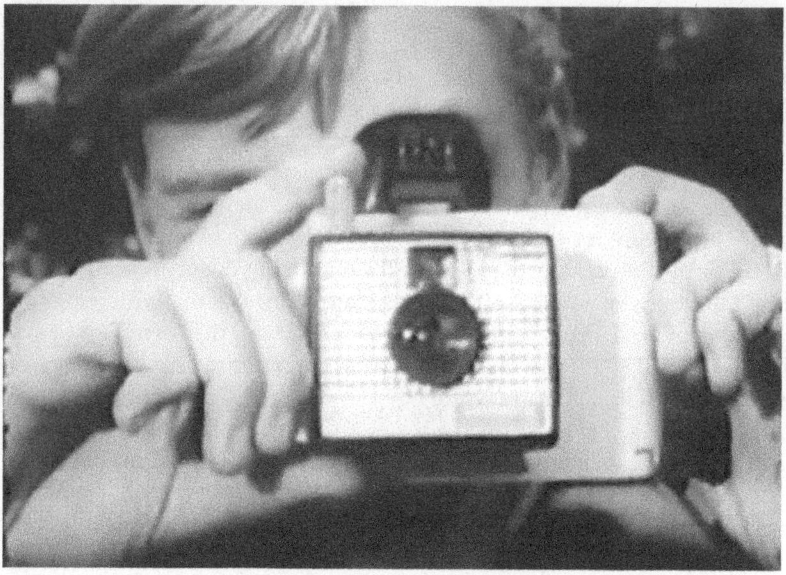

Figure 7.3 Still from the Polaroid Swinger TV ad, 1965.

from a harried off-stage observer into someone who was a natural part of the event."[17] For Land, the SX-70 was the culmination of the instant-image experience, one that had the potential to bring together people in new relationships – a kind of techno-optimism that intersected with the happening culture of the 1960s and 1970s. In a small booklet he wrote in 1974, he states:

> [A] new kind of relationship between people in groups is brought into being by SX-70 when the members of a group are photographing and being photographed and sharing the photographs: it turns out that buried within us – God knows beneath how many pregenital and Freudian and Calvinistic strata – there is latent interest in each other; there is tenderness, curiosity, excitement, affection, companionability and humor. . . . We have a prehistoric tribal competence for a non-physical, non-emotional, non-sexual satisfaction in being partners in the lonely exploration of a once-empty planet.[18]

Land may have wanted to couch the instant image in a vocabulary of human consciousness and spirituality, but the Polaroid instant camera was much more about hipster youth culture than it was about finding spiritual connection. The instant image is not only hip, it is also about sex – sold by the company not directly so much as in veiled terms, with the use of code words such as "intimacy."[19] The Polaroid is a more private image – no need to send off those potentially embarrassing images to a photo lab, they can stay within the home (or the party). So the equation of the Polaroid with sex and titillating images (instant ones at that) happened early on.

Artists began to experiment with Polaroid images from the beginning and took to the SX-70 rapidly; many used multiple SX-70 images to create collages (notably David Hockney and Chuck Close) and to experiment with instant images (Robert Mapplethorpe, Andy Warhol, William Wegman) and rock musicians such as The Talking Heads used SX-70s on album covers. Polaroid began early on to acquire a corporate collection of photographs, with Land collaborating with Ansel Adams to choose work, and then over the years provided free film to photographers in exchange for some of their work. Polaroid would develop a large-scale 20 (multi) 24 instant camera in 1974, out of Land's desire to create a larger Polaroid, and these cameras would become legendary with artists for their very distinctive physical quality and resolution. Five of these cameras were distributed to urban centers around the world, such as New York, Prague, and San Francisco, and created ad-hoc art Polaroid studio laboratories where artists would come to produce images.[20]

Ironically, while Kodak achieved its brand dominance by selling an equation of photography and nostalgia, Polaroid has spawned a nostalgia culture for itself – for the instant image in the form of the white-bordered SX-70. A project in the Netherlands, led by former Polaroid employees, called The Impossible Project, resurrected instant analog film and created a kind of subculture of Polaroid image-makers who share images on its website.[21]

Facebook

Now it's 2015. Kodak is bankrupt, attempting with its intellectual property of patents to reinvent itself as a smaller company. The Kodak Theater of the Academy of Motion Picture Arts and Sciences in Los Angeles (home of the Oscars Ceremony) has been renamed the Dolby Theater. Polaroid is twice bankrupt, and its reinvention, after being sold a few times, is in doubt. Yet the cultures each spawned are now embodied in the practices of digital images. The razor-blade business strategy that created an overdependency on film in both these companies has been transformed into the digital photograph *platform*. With the rise of the mobile-phone camera, the practices of amateur photography have been dramatically transformed. In this new, rapidly changing realm, Facebook and Google (among a host of other companies) aim to establish themselves as the primary platforms through which user-consumers will self-document and share images.

This shift has been characterized by cultural commentators and the technology press as the death of the analog photographic-film industry, replaced by social media as the future of photography. What are the implications of the transference of a set of photographic practices from the analog photograph to not only digital imaging but also to social media and smart phones? Certainly, one could argue that smart phones and mobile-phone cameras are increasingly making digital cameras obsolete. The mobile-phone camera changes photographic practices through two very specific features: first, users tend to carry them with them all the time, thus dramatically increasing camera availability, which has the immediate impact of increasing the number of photographs taken; and, second, the linkage of the phones to websites, blogs, Facebook accounts, and other social-media venues means that images taken on mobile phones can be instantly shared. In this context, we could easily say that the concept of the Kodak moment is increasingly obsolete, as mobile-phone cameras encourage constant self-documentation and photographing rather than the unique photo moment. The Kodak moment was unique, planned for, and orchestrated for the camera as a means to produce ritualized memories. The mobile-phone image is, by contrast, the instant image that is shared in an accumulation of images.

The "effortless abundance" initiated by Kodak is thus taken to new levels in this context of digital-photographic practices. As technology writer Alexis Madrigal has written, "what could describe our current world of shareable digital photography better than effortless abundance?"[22] The ease with which large numbers of images can be produced, uploaded, and shared (if not instantly "curated") with minimal effort has shifted the demands of life documentation. Documenting one's life has been reshaped as a practice that constitutes a much more time-consuming and demanding process – one is asked to photograph every day, to share images constantly, to "update" regularly, and to curate one's own mundane moments for one's friends (and Facebook "friends"). Hence, there have emerged numerous webblogs and Facebook walls that are about ritualistically photographing everyday activities (such as documenting meals, logging hotel rooms, etc.). Importantly, then, the effortless abundance of nineteenth-century early domestic photography is transformed into a vast output of images of the mundane, if not images that celebrate the mundane. Here, we can see the shift away from the Kodak "moment" that was special not only because it was a ritual or a moment of social gathering but also because it did not happen very often.

The new set of digital and mobile photographic practices are not only about abundance, mobility, and accumulations of the mundane; they are also about deploying pastiche to play with the styles of photography throughout history. Thus, contemporary imaging-making practices include a lot of play with historical image styles in the creation of faux historical images. Nowhere is this clearer than it is with the enormously popular Instagram. Its very name referencing (in postmodern ironic fashion) both the telegram and the Kodak Instamatic, Instagram quite self-consciously charts its ethos to the legacy of Polaroid. Indeed, its logo is a facsimile of the Rainbow SX-70 OneStep Land Camera (originally released in 1978). Instagram states its primary aim as a mobile-phone application is to compensate for the poor quality of most smart-phone photographs by using filters to make them look better and by allowing users to share their images on multiple platforms with ease. Thus, in the short time since the app has been available, Instagram has acquired millions of users and its signature style of a sepia or faded-color image (called 1977), or its SX-70-like framing white border. Hipstamatic, its competitor, sells itself with the slogan, "digital photography never looked so analog."

It is thus not surprising that these new photo apps have also produced a culture of nostalgia in which huge numbers of digital images circulate via social media that look like aged snapshots. For many users, the faded color of photographs from several decades ago, or the faded sepia tones of black-and-white images, evoke an immediate sense of loss and desire to dial back the clock. The instant-nostalgia style that is created by Instagram, for instance, turns even banal images into something immediately meaningful, resonant, and evocative. These images enfold the present into the past, giving contemporary snapshots the constructed style of the relic, the album, the held-on-to image. As Nathan Jurgenson has written, "the faux-vintage photo is an attempt to create a sort of 'nostalgia for the present,' an attempt to make our photos seem more important, substantial, and real. We want to endow the powerful feelings associated with nostalgia to our lives in the present."[23]

Pastiche in the context of social-media photography also extends to the highly networked relationships that social-media sites encourage. Thus, for many users, in particular those who are older, Facebook can constitute a kind of continuous reunion, encouraging users to find former high-school and college classmates and connecting users via past experiences. As Susan Dominus writes:

> In its early days, media pundits hailed Facebook as the social application of the future, and yet what it really does is change our relationship to the past. Facebook makes contact so casual that it allows people to leapfrog back instantly to a former you, one you thought you had left behind—maybe one you had worked hard to put firmly in the past.[24]

The practice of photo tagging on Facebook and rituals such as TBT ("Throwback Thursday") can also have the effect of pulling users (willingly or unwillingly) into their pasts through the nostalgic posting of older photographs. This sense of simultaneously inhabiting several selves from one's life (for one's Facebook friends from high school, for instance, one may be frozen in that former persona) is a common response to the Facebook experience.

Behind these practices lies the notion that we are curating our activities for future viewers, including ourselves. Just as the photograph evokes mortality (a future death) and continuity (future viewers of the images), the documentation practices of social media affirm for users that they have a future that they are documenting toward. The accumulation of images that creates a personal database of one's life on social-media sites such as Facebook, has the effect of increasing the stakes of users' relationship to social media, and of equating activity with connectivity.

Rob Horning argues that Facebook is emblematic of the rise of the "data self" that emerges after one generates a sufficient quantity of online data. He writes:

> And though Facebook wants "the Timeline to be a place for self-expression: A way for users to reveal who they are and what their lives are about," it has provided a tightly controlled and highly formatted medium for it that emphasizes standardization. . . . It imposes the metaphor of life and memory as a stream, which . . . is not some natural, neutral reflection of how we remember but a reshaping of life into narrative, which suits Facebook's ends. The more work we put into making a coherent story out of the data Facebook collects, the more useful, marketable information we give them.[25]

Thus, in this context, selfhood is only derived through documentation and the accumulation of data – it is not self if it is not a part of the abundance of posted updates and images. A typical young user of Facebook (such as a high-school student or a college undergraduate) has many thousands of images on their Facebook Timeline. The incitement to constantly update (often embodied in the constant changing of one's Facebook profile thumbnail image) can be traced to a sense that one is constantly changing, and that digital practices demand vigilance and constant activity. As Susan Murray has written, this means that digital photography signifies *transience* much more than the photographic image, which has long been understood to signify loss.[26] She writes that "there is an implicit acknowledgement of the inability of photos to hold onto certain moments. Rather than interpreting this as a type of death, in the display of digital photography in social network sites (and photoblogging sites) there is an already accepted temporariness to one's sense of publicly present self in all of life." The self cannot be established by the rituals of Kodak moments; it is ever moving, ever changing, ever in need to the affirmation of an update. The *update* is its primary mode.

This transience of the self (the self in need of constant updating) can be seen not as a kind of remaking of the photo album but in fact about the importance of the concept and valuing of *sharing* to notions of the self. Thus, the self does not exist alone, but rather only through sharing. It's worth noting that while photographic albums are often understood as that most precious of family objects, the first household item grabbed in moments of disaster, most studies show that people rarely look at them. As Don Slater has written, the family album "in a concrete or metaphoric sense – is hyper-valued yet plays little part in everyday life. . . . We need to know they are there (and in a persistently existential sense) but they are not part of the everyday practices which involve images."[27] We have only to reflect on the common stereotype that for non-family and family members alike, the emergence of the family album or slide show is often greeted with dread and seen as an ordeal to suffer through. Thus, there seems to be little valuation of the idea of sharing the family album within the everyday.

Yet sharing is the essence of social media, not only in its practices but also in the business strategies of social-media companies – they generate income through the sharing practices of their users. The photo in social media is valued primarily because it is shared and thus helps to affirm information networks that provide avenues of brand messages and the reservoirs for consumer profiles. This redefinition of the practice of photographic documentation has implications both for consumer privacy and for the ways in which consumer data have become the essence of postindustrial capitalization (of which social media is perhaps quintessential at this moment in its development). Put simply, if Facebook is becoming the platform for personal photographic histories, it is, by extension, the owner of those photographs and Timelines. It establishes the rules by which users construct and share those photographs, and it guides the practices by which these images are circulated through social networks. Facebook thus stands (with its more than one billion users) to play the dominant role once played by Kodak to actually guide cultural practices of photography. Its corporate values are thus crucial to the shaping of photographic practices as linkage, network, and sharing, even extending to sharing beyond personal networks.

Much has been made of Facebook's shifting privacy policies and the challenges of what networked privacy might mean.[28] Helen Nissenbaum has argued that rather than understanding privacy in terms of control and access, we must look at privacy in context in order to understand how contemporary digital-media practices, for instance, produce claims of privacy violation and the extent to which these are legitimate. In arguing for a concept of contextual integrity, through which expectations about privacy can be framed and understood, Nissenbaum states that we must use contextual analysis as a means to move beyond simple notions of public/private and categories of "sensitive information."[29] The personal photograph would seem to constitute a particular kind of challenge to changing concepts of privacy, in particular in relation to the capacity of sites such as Facebook to retain the content of personal pages.

Yet, it is on the question of the platform that Facebook's aim to establish itself as the photographic album of the future raises a set of paradoxical concerns. Here, the replication of the razor-blade strategy in the model of the platform is crucial. Christian Fuchs writes that the business model of a corporation such as Facebook produces surplus value not only through its programmers and marketers, but also through the users and "prosumers" (producer-consumers) who produce its user-generated content. He writes:

> A widely-used accumulation strategy is to give the users free access to services and platforms, let them produce content, and to accumulate a mass of prosumers that are sold as a commodity to third-party advertisers. No product is sold to the users; the users are sold as a commodity to advertisers. The more users a platform claims, the higher the advertising rates.[30]

In this context, sharing is converted to capital and the photographic practices being nurtured by the Facebook platform – linking, circulating, posting, tagging – form the foundation of the Facebook business strategy. As I noted before, the Facebook platform follows from the Instamatic or SX-70 camera as a product *portal*, establishing an affective relationship between the brand and the consumer. In social media, however, the value is shifted from film to consumer data.

Finally, though, if we look at these business histories and photographic practices together, we can see that there may be many reasons to wonder about the precariousness of our digital-photo albums. Facebook retains a dominance in the realm of social media by presenting itself as the primary portal for all social interaction, the means through which users will connect to their networks, share their images, and construct their identities. Yet, looking at the business histories of Kodak and Polaroid should remind us that most companies cannot dominate markets for very long, that the features of market dominance make it hard for them to see outside their bubbles. Kodak and Polaroid could not see outside the bubble of dependence on the photographic-film market, and Facebook cannot see a world in which it is not the only or the most dominant platform. But history tells us that Facebook's world is transient; its model for photography will be, inevitably, subject to change.

Notes

1 Kevin Kelleher, "Face.com: Facebook Reinvents the Kodak Moment," June 19, 2012, http://readwrite.com/2012/06/19/facecom-facebook-reinvents-the-kodak-moment.

2 See Susan Danly, Chris Thompson, Judy Glickman, and Irving Ellis, *For the Love of It! The Photographs of Irving Bennett Ellis* (N.p.: Ellis Press, 2008).

3 Associated Press, "Number of Active Users of Facebook Over the Years," January 30, 2013, Yahoo News, http://news.yahoo.com/number-active-users-facebook-over-230449748.html.

4 Nick Bilton, "Disruptions: Innovation Isn't Easy, Especially Midstream," *New York Times*, April 15, 2012.

5 *The Economist*, "The Last Kodak Moment?" January 14, 2012.

6 Quentin Hardy, "At Kodak, Clinging to a Future Beyond Film," *New York Times*, March 20, 2015.

7 See, for instance, Alecia Swasy, *Changing Focus: Kodak and the Battle to Save a Great American Company* (New York: Times Business, 1997).
8 Unlike Kodak, Fujifilm managed to survive the transition to the digital by becoming a more flexible and diversified company, deploying its expertise in photo-sensitive chemicals into cosmetics, medical imaging, and it retains a significant percentage of the digital-photo kiosk business. See K.N.C. (Schumpeter), "Sharper Focus: How Fujifilm Survived," *The Economist*, January 18, 2012, www.economist.com/blogs/schumpeter/2012/01/how-fujifilm-survived.
9 Nancy Martha West, *Kodak and the Lens of Nostalgia* (Charlottesville: University of Virginia Press, 2000), 206.
10 Ibid.
11 Ibid., 2.
12 Ibid.
13 Richard Chalfen, *Shapshot: Versions of Life* (Bowling Green, O: Bowling Green State University Popular Press, 1987), 10.
14 West, *Kodak and the Lens of Nostalgia*, 183.
15 Peter Buse, "Polaroid into Digital: Technology, Cultural Form, and the Social Practices of Snapshot Photography," *Continuum: Journal of Media & Cultural Studies* 24, no. 2 (April 2010), 218.
16 Christopher Bonanos, *Instant: The Story of Polaroid* (New York: Princeton Architectural Press, 2012), 75.
17 It can be seen at www.youtube.com/watch?v=5jaiq_ZZ_eM.
18 Cited in Bonanos, *Instant*, 105.
19 Ibid., 71.
20 Ibid., 79.
21 www.the-impossible-project.com/.
22 Alexis Madrigal, "The Triumph of Kodakery: The Camera Maker May Die, But the Culture It Created Survives," *The Atlantic*, January 6, 2012, www.theatlantic.com/technology/archive/2012/01/the-triumph-of-kodakery-the-camera-maker-may-die-but-the-culture-it-created-survives/250952/.
23 Nathan Jursenson, "The Faux-Vintage Photo," *Cyborgology*, May 14, 2011 http://thesocietypages. org/cyborgology/2011/05/14/the-faux-vintage-photo-full-essay-parts-i-ii- and-iii/.
24 Susan Dominus, "Can Facebook Be Your Friend?" *Real Simple Magazine* (April 2009): 159, www.realsimple.com/work-life/technology/communication-etiquette/facebook-friend.
25 Rob Horning, "The Rise of the Data Self," PopMatters, January 25, 2012, www.popmatters.com/pm/post/153721-/. See also Rob Horning, "Facebook in the Age of Facebook," *The New Inquiry*, April 19, 2012, http://thenewinquiry.com/essays/facebook-in-the-age-of-facebook/.
26 Susan Murray, "New Media and Vernacular Photography: Revisiting Flickr," in *The Photographic Image in Digital Culture. Second Edition*, ed. Martin Lister (New York: Routledge, 2013), 165–182.
27 Don Slater, "Domestic Photography and Digital Culture," in *The Photographic Image in Digital Culture*, ed. Martin Lister (London: Routledge, 1995), 139.
28 danah boyd, "Networked Privacy," *Surveillance & Society* 10, nos. 3/4: 348–350, www.surveillance-and-society.org. See also danah boyd and Eszter Hargittai,

"Facebook Privacy Settings: Who Cares?" *First Monday* 15, no. 8 (August 2, 2010), http://firstmonday.org/htbin/cgiwrap/bin/ojs/index.php/fm/article/view/3086/2589.

29 Helen Nissenbaum, *Privacy in Context: Technology, Policy, and the Integrity of Social Life* (Stanford, Calif.: Stanford University Press, 2010), 232.

30 Christian Fuchs, "The Political Economy of Privacy on Facebook," *Television & New Media* 13, no. 2 (2012): 144.

Chapter 8

Forgiving without forgetting

Contending with digital memory

Ira Wagman

One of the achievements of the "information revolution" is the Internet's increased capacity to record, store, and disseminate embarrassing things about people. Indeed, the Internet abounds with videos of people doing embarrassing things, often without their knowledge. We are tagged in photos that end up on Facebook that may make us cringe. Things marked as "#fail" highlight attempts that have missed their marks. At one end of the scale we have "bloopers" of various kinds, while at the other end the situations are much more serious. Here, photographs of genitalia intended for a single viewer end up circulating online, tricks or dares become data, or a night of drunkenness ends up on Facebook. In between, we have a range of social behaviors, such as slips of the tongue, inappropriate asides, leering stares, or lapses in judgment that we are unable to take back. The popularity of virtual private networks that mask the IP address of your computer, such as Tor or Hotspot Shield, or the "private browsing" option on Firefox, does not stem entirely from a concern over government spying but as a kind of insurance against embarrassment, a way to protect ourselves from having our online forays made public to our various social networks.

Feelings of embarrassment, guilt, fear, and shame are prominent emotional characteristics of living in and through media technologies, what Mark Deuze has characterized as "media life."[1] In this chapter I want to think about online embarrassments as way to explore ethical questions associated with interpersonal relations – how are we to live with each other against the backdrop of the presence of embarrassing things and persistent concern about the management of future embarrassments?

In a recent essay Luke Purshouse explained that embarrassment "seems to occur only in situations where a subject views himself as actually coming into contact with other human beings." It may also arise, he explained, from "imagining situations involving contact with other people, rather than really being in one."[2] This neatly explains some of present concerns about online privacy, namely that information about us will return at some future point to haunt us, whether that is true or not. It is perhaps not surprising, then, that these kinds of acts lie at the center of contemporary concerns about the ways we manage ourselves online. danah boyd refers to this as "identity work in networked publics" in her study of the online behavior of teenagers[3], an extension of Erving Goffman's sociological research on the presentation and maintenance of self in face-to-face encounters. We may have long been concerned with the presentation of self in spite of the fact that we can never know what others really think of us. What may be different about the current moment, however, is our belief that we have the facility to control those perceptions more than in the past because we now possess the means to produce more representations of ourselves than before.

We might understand this concern with digital memory as the product of the decline of what Andrew Hoskins has perceptively called "decay time," in that information stored on recordable media would inevitably suffer and wither away through physical decline that would slowly deteriorate individual and social memory. Such "familiar deterioration," as he called it is now not possible in a world characterized by connectivity, the easy capacity to make high-quality copies of images, texts, and sounds, and the incredible storage capacity of individual computers and large server systems.[4] In what follows in this chapter I address the ways we have tended to date to deal with this issue. To my mind there are two overwhelming tendencies: one associated with managing; the other associated with deleting. The first concerns the management of online profiles in the name of protecting one's reputation; the other has to with giving people the power to delete information once widely available on the Internet. Both of these concepts converge through recent developments associated with new privacy laws being debated in Europe giving people "the right to forget."

In reviewing these two examples I cannot help but wonder if something is missing in thinking in terms of management and deletion as coping mechanisms, or that they create bigger problems than they were supposed to resolve, and serve mainly as technical solutions to technical problems about the relations between humans and computers. However, as Ori Schwarz has shown, what is truly needed is an understanding of memory that moves beyond this kind of thinking and toward "a language of interobjective sociality in which humans and non-humans encounter each other and work on each other."[5] My contribution toward developing that language is to think about the possible dividends that may come from a consideration of forgiveness. To put it bluntly, I believe that the ethical challenge may well be about how we let others "off the hook" about the embarrassing things they may do online rather than whether we should devise better and more elaborate ways to delete information from the Internet. Might there be something to be gained by thinking ethically in terms that would permit forgiveness without forgetting, of managing without redacting, or of accepting without deleting?

A burp in public

Permit me to offer a painfully simple example: I am standing next to you in a busy bar. We are both enjoying an *aperitif*. Then, unexpectedly, the spirit of "happy hour" is upset because I have just burped out loud. I am horribly embarrassed about this on my own account, but also I have clearly offended you, judging from the look on your face. Realizing what I have done I turn to you and offer an apology. I hope that you believe my sentiments are genuine, and you tell me so. We then go back to our conversations with friends and the evening continues without further incident.

Now let us add something different to the mix. This time you are shooting a video of your friends in the bar and happen to catch the act. You stop filming to listen to my apology and you are gracious enough to say all is forgotten. However, hours later you then turn that burp into six-second video on Vine and then post a longer version of that video on YouTube with the title "Loser Burps in Bar" that attracts a decent number of hits, likes on Facebook, and re-tweets. It is through one of those re-tweets that I discover what has happened. Now I have to decide what to do about it. I can try to contact the person who posted the video to take it down. I may try to contact YouTube and Vine to "report a problem." If I am extremely embarrassed, I may even seek legal action to have the video removed.

There are some obvious differences here. In the first case, the situation is resolved locally; the apology is offered in context and the person who grants the apology does so in real time, if you will excuse that term. In the second, the situation may be solved locally but the incident circulates globally. It is offered without context and it is highly likely that the viewers will never know if an apology has been granted. This may be a matter of degree; even if the apology was granted in the first case, you may choose to recount the story to another friend the next day at the office, selectively providing the details you want to offer for maximum comedic effect. What is also different is that my embarrassment is now something I believe that I have to *manage* through a series of procedural moves involving contacting strangers and seeking legal actions. The remedial work has become a remedial *career*.

We might describe the situation I have described here as a function in Roger Silverstone's terms, as the erosion of what he called "proper distance." For Silverstone, proper distance "refers to the importance of understanding the more or less precise degree of proximity required in our mediated interrelationships if we are to create and sustain a sense of the other sufficient not just for reciprocity but for a duty of care, obligation and responsibility."[6] Silverstone argued that media technologies have played a contradictory role. On the one hand, they offer the promise of bringing people into proximity with each other. On the other hand, they increase the sense of strangeness between them. A part of that erosion of proper distance is the fact that new communication platforms are efficient, relatively inexpensive, and incredibly easy to use. This may be an obvious fact to the reader but it bears repeating here. Sites such as Facebook are wonders of design; one can transmit information and status updates without even the slightest knowledge of computer programming. One forgets that a key prerequisite for early digital culture was knowledge of HTML or Java. This was once a technological obstacle that slowed down processes of "lifestreaming." You could not update your status unless your webmaster could put it online for you – try to remember what it used to be like to update a webpage for a sense of what things were like. However, these matters are now restricted to more traditional websites, whereas sites such as Twitter and Facebook enable instant communication without programming knowledge. As José van Dijck put it, the digitization and easy manipulation of photographs and video permit endless pathways for distribution and practices of what she calls "unintended repurposing."[7]

However, another feature worth mentioning here is the virtual impossibility for online communication to be taken back. Taking things back – whether in person or in writing – represents a break to speeded-up communication. The first thing that comes out of our mouth may not be the right thing or the most appropriate thing. Gmail provides users with five seconds to retrieve an email; it is remarkable how few other platforms provide even that basic capability. Indeed, one wonders if not having a retrieve function on platforms such as Twitter represents the company's secret desire to court the slip-up, to turn your faux pas into things that go viral. New applications attempt to capitalize on this sentiment; Snapchat falsely promised its users that they could send photographs to others that would automatically "self-destruct" and disappear from their online memory. However, such applications actually underscore the current models of online information management, as the photograph does not actually "disappear" but rather finds itself stored on the company's servers.

What has emerged are a number of "micro-celebrities," to use Alice Marwick's term,[8] who symbolically stand-in for the embarrassments of the Internet. In one celebrated case the public-relations executive Justine Sacco was fired from her job after a racist tweet about AIDS in Africa before boarding a flight to Cape Town.[9] In Viktor Mayer-Schönberger's book, *Delete*, the author draws upon the case of Stacy Snyder, an education student who was denied her teaching certificate because she had posted a picture drinking out of a plastic cup and wearing a pirate's hat on her Myspace page because it exposed unprofessional behavior.[10] In Canada there was the case of Natalie Blanchard, a Canadian employed by IBM who took sick leave to address issues surrounding her depression. She was told by her doctor to take a vacation and to seek more sunlight, a recommendation taken by many Canadians in the deep, dark throes of winter. Blanchard then posted pictures of herself in a bikini on the beach as well as attending a male strip show. According to news reports, Blanchard's insurance company discovered the photographs and, convinced that she was no longer ill, stopped making her benefit payments.[11]

Finally, on its "Currency" blog, *The New Yorker* recently ran a story about hundreds of lawsuits launched by the erotica website x-art.com to surfers for copyright infringements, which were often settled out of court because of the fear that the individual's online predilections would be made public.[12]

In a sense, these contemporary cases represent the coming together of longstanding issues present in previous media forms: the letter ends up in the wrong hands; something we have said to the press has been taken out of context because it does not fit into "sound bites"; or we are recorded saying something that wasn't intended for public consumption. Of course, in these cases the idea is that the message will be not understood out of context, that one will only hear or see one aspect of "total" message and miss the connection between the utterance and its object.

At another level, though, and as John Durham Peters has observed, this is in large part a function of the inherently leaky nature of communication: "In a face-to-face setting talk may be overheard by unauthorized listeners (eavesdropping), be passed along to third parties (breach of confidence) or fail to hit its desired target (misunderstanding)."[13] The increased capacity for communication made possible by digital technologies along with the increased quantity of those communicative exchanges that can be saved by storage devices by its very nature also multiplies the prospects for leaks. The more people taking photographs, the more photographs being taken, and the more photographs that are saved simply increases the prospects that more things will "get out" beyond our control. Since those leaks are also now part of databases, and their appearance may be determined by algorithms, memories now materialize and rematerialize in random and unpredictable ways depending on whatever is entered into a search engine.[14] As true as this may be, it is perhaps more accurate to say that the "memory status" of individual media production (i.e. your photos) is now treated in much the same way as the media artifacts of mass-media organizations, such as the video libraries of television networks, or companies that provide stock photography.[15] Here, an event, program, or place can be recorded, copied, coded, and cataloged in a number of ways for future re-appearances in newscasts, retrospectives, or for commercial purposes to suit the specific needs of clients. It should perhaps not come as a surprise, then, that the remedies associated with the management of online embarrassments derive from practices that emerged out of the dissemination of communication out of context to audiences with loose bonds to the speaker: public relations.

Managing mediated embarrassments

In an article published in *Public Relations Review* in 2001, scholars discussed the emergent practice of "reputation management." In their discussion Hutton et al. note that the very idea of reputation management itself represents a way of making sense of some things that are not particularly well defined. "Concepts such as 'image' and 'reputation' are not generally something that can be managed directly," the authors write, "but are omnipresent and the global result of the firm's behavior." Attempting to manage one's reputation, they write, "might be likened to trying to manage one's own popularity (a rather awkward, superficial, and self-defeating endeavor)."[16]

A number of firms have emerged to provide that very service. One of the leaders in the marketplace for "online reputation management" is reputation.com. The company's website offers customers the capacity to suppress negative search results, uncover and protect personal data that circulate online, and to get better, authentic, and more original reviews from customers. All of these efforts come as a reflection of the company's belief that "controlling your online reputation and private data is key to success in our digital world."[17]

As Nora Draper has explained, online reputation management companies emerge in large part out of the tension between existing legal models, which control speech acts for media corporations such as slander and libel, and the free-speech ideals, which are part of the ethos of the Internet. They also operate on an important assumption about time. Bringing legal actions against irresponsible speech is a time-consuming process; however, the management of online activities is something that has to happen quickly. In her analysis, Draper offers that in the name of providing timely remedies to online problems, such services also tend to be costly, lack due process and regulatory oversight, and support a culture of visibility by encouraging the distribution of correct information, not to put a stop to the putting out of information in the first place. More than that, Draper argues, the existence of an online reputation management industry itself is indicative of the central problematic, which is "if you don't take control of your online identity, someone else will do it for you."[18]

A company such as reputation.com is aimed at protecting the integrity of high-status figures such as doctors and judges but, of course, since more and more things are done online, and since more people are engaged in online activities, there is much more at stake in managing aspects of your online personality. An example is the company AIM Sports Reputation Management, based in Atlanta, Georgia. The company was started by a public-relations firm, Porter Novelli, to cater specifically to "college and professional athletic programs with assistance in leadership, ethics, and brand management." One of the company's senior executives is Bill Curry, a former college coach and athletic director. Of particular interest to the firm is its specialty in helping organizations avoid "reputational harm" that can come from controversies brought on by players or by management officials. Among the services offered by AIM include the usual public-relations fare such as crisis management, media training, and strategic communication, but also "social media auditing," which monitors the online activities of players and fans, and keeps organizations abreast of what players may be tweeting or what individuals may be saying about individual players or the team itself.[19]

The point here is not to suggest that those who offer or use reputation management services are wasting their time or are working from poor assumptions. For our purposes, though, their presence reflects a few important assumptions about digital memory. First, it reveals that reputation is something that can or should be managed in part because of the omnipresence of communication. Second, it extends the idea that communication is risky. Finally, it brings us into a temporal logic that failing to act now to minimize that risk brings with it significant consequences in the future. These powerful assumptions resonate despite our inability to truly know what our reputations are; such firms likely act in the name of the user's self-conception of his or her own reputation as much as the version circulating on, and presented by, the Internet. We can only guess what other people, including strangers, casual acquaintances, and even loved ones will really think of us.

Right to forget?

The front page of the May 13, 2014 edition of the *New York Times* announced that the European Court of Justice came down with a judgment against Google suggesting that the search engine must now erase links to sites that may make some people uncomfortable, an extension of what is commonly called "the right to forget."[20] The justices in Luxembourg ruled in favor of Spanish lawyer Mario Costeja, who brought the case. The judgment cannot be appealed, and now it is up to the legal systems of the European Union (EU) member countries to apply the laws as they see fit. The news from the courts came as the European Parliament is in the process of updating its data production laws to build in similar rights as articulated by the courts. These ideas are reflected in the EU's Charter of Fundamental Rights, which states that "Everyone has the right of access to data which has been collected concerning him," the decision reads, "and the right to have it rectified."[21]

At the heart of the proceedings was a 36-word announcement in a Catalonian newspaper that announced that Costeja's house was being repossessed in order to pay off his debts. The notice appeared among other notices of a similar nature in the paper as part of its public announcements. The notice appeared in 1998, but it continued to reappear in Google searches. Convinced that this news would follow him forever, he argued that Google and other search engines should exercise his right to put the past behind him and bury the links to the newspaper notices.

In one sense, we might say that the decision attempts to rebalance information flows that take individual rights into consideration. It is significant, I think, that the decision comes against an American company operating in Europe. Indeed, the affairs of US high-technology firms in Europe are a particular concern of the EU and its member-states. Such moves come on the heels of previous efforts by a number of European countries to extend privacy rights to a range of online applications. Germany is arguably the most notable of those countries: it has imposed fines and restrictions on Google's Street View service for breaches of privacy. In 2013, a federal court was successful in getting the company to remove the algorithm of its "autocomplete" function that attached words such as "fraud" and "scientology" to an anonymous individual.[22] The EU may be a fractious place that mixes bureaucrats and ultra-nationalists, yet it would seem keeping Google in line and working against future embarrassments may be something that unites those different positions.

A significant influence on "right to forget" policies is the work on memory in the digital age that comes from Viktor Mayer-Schönberger. His book *Delete*, mentioned earlier, takes as its starting position that the incredible powers of the Internet are changing what he calls "our default settings" that make it possible for societies to forget. Among the solutions offered by Mayer-Schönberger is that information stored on our various devices would have an expiry date. Once that date is reached the information would automatically be deleted. For Mayer-Schönberger, this is a technical solution to the Internet's capacity for perfect memory and as an efficient means to "re-introduce forgetting." Such a measure is "designed to confront us with (and thus remind us of) the 'finiteness of information' – in other words, that information is inexorably linked to a period in time and that over time most information loses its informational value, much like yesterday's newspaper, or an old joke we have heard too many times."[23]

Although it is certainly the case that the Internet captures many things, the memory collected on the Internet is by no means more "perfect" than any other medium that preceded it. It may be more voluminous, more connected, more global, and more elaborate, but it is certainly not *perfect*. What may be more precise to say that its discussions hinge upon is precisely the social problem of the extent to which we will be able to forgive in the absence of policies that bring about forgetting. One would have to accept the argument that a past experience – such as a bankruptcy – would necessarily result in something that will constantly follow the person for the rest of his or her life. Mayer-Schönberger also fails to account for the ways that mediated memory is already part of the processing of a kind of forgetting. As José van Dijck points out, we have long used photographs from past events to prioritize remembering some things over others. Indeed, photography is part of the way we "re-model our self-image to fit to pictures taken at previous moments in time."[24] In addition, van Dijck reminds us that we have long wished to manipulate our images of ourselves by shooting at particular angles, wearing flattering clothes, and applying chemical "filters" to make us look better. Control over images then, represents "a balancing act in which photographic images 'enculturate' personal identity."[25]

We can take a small measure of the ways in which the notion of "forgetting" is conceived at the level of the EU by considering the ways that the new proposed law was recently communicated to the public at large. In this case we can see how the EU has decided to frame the issue of privacy, but in doing so also how it frames a particular conception of the Internet. The theme of nakedness is prominent in both a public-service announcement posted online and a number of cartoons posted on the Ministry of Justice's "mini-site" of data-protection rules. In each case the Internet surfer is presented as sitting at the computer fully clothed, only to discover with a few keystrokes that they are now stripped naked.[26] In the public-service announcement, the surfers are not aware of their nakedness, continuing on to board a plane or meet a friend. In the cartoons on the mini-site, embarrassment is the central theme, with images of a surfer coming into an Internet café with clothes and leaving without them, or of people realizing their data has been now taken over by pirates, or of evil machines tempting customers into giving personal information, such as an address.

In her recent characterization of the proposed law, the EU's Commissioner for Information, Viviane Reding, has on a couple of occasions drawn attention to the plight of young people, suggesting that the law is needed in order to ensure that people who might do things as youngsters that they would regret will now have the chance to pursue a full and fulsome life free of routine reminders of past mistakes.[27] Reding's characterization of children locates the current discussion on forgetting within other examples of moral panics directed at young people, stretching from movies and comic books to alcohol and marijuana. These examples all perpetuate longstanding perceptions about media technologies as autonomous, and of users as unwitting dupes.

We might view these policies as marking the extension of a certain logic about nationalisms and expanded to a regional or continental scale. As Benedict Anderson has written, channeling the French philosopher Ernest Renan, new nationalisms in Europe emerged as "awakening from sleep," with a conscious attempt at forgetting previous tragedies that might mark out tensions between ethnic and religious groups. However, for Anderson there is a powerful temporal idea here at play: "Awareness of being imbedded in secular, serial, time, with all of its implications of continuity, yet of 'forgetting' the experience of this continuity . . . engenders the need for a narrative of identity."[28]

Many of the most recent developments regarding online memory have been to repatriate the data within specific countries. If servers collect information about German citizens, so goes the thinking, they should be stored in Germany and subjected to German privacy laws. At the heart of "the right to forget" – our previously poor Catalonian, Mario Costeja – is the idea that the newspaper that published the announcement is not expected to remove the physical copies of the papers or even to take the information down from the newspaper site, as this would be in the eyes of the judicial decision "a falsification of history."[29] What the "right to forget" directives appear to do, however, is to limit the circulation of that embarrassing data. One could continue to find the paper copy of the newspaper at a local library or in an institutional archive, for example, but it will not be available online. Of course, one might think that the opposite solution would be true; if forgetting and deleting policies were to be indifferently applied it would be better if they applied locally (say Catalonia) where the individual's future is more likely at stake than globally. However, the decision operates on two levels by managing the flow of information deemed to be embarrassing and by reinstating the importance of nations and the EU itself as the region in which such things will circulate. Such a move reproduces key notions in the EU about mobility within defined, albeit much larger, spaces. As a region that started as an economic partnership, then, it is significant that a bankruptcy serves as the tipping point for new rules on privacy.

Distant forgiveness?

From this cursory account we can see a few problems emerging. The solutions being offered by reputation management firms and through forgetting rights emerge largely out of a concern with new media's capacity for embarrassment. From there, they treat those problems either by outsourcing them to other companies or by conceiving of online memory in terms that speak in a voice that prizes individual rights but in the end underwrites particular European rights. Arguments for deletion, such as those of Mayer-Schönberger, prize efficiency and perpetuate myths of perfect memory to underwrite claims for deletion in the public interest. Neither appears, then, to represent a satisfactory solution to the problem. With the remainder of this chapter I want to suggest that a more robust thinking through of the concept of forgiveness may be in order precisely because it offers a way through to make sense of the link between online memory and embarrassment.

As with notions such as embarrassment and reputation, forgiveness is an equally tricky term. Christel Fricke notes the role of forgiveness in religious practice, such as in the Lord's Prayer. However, Fricke notes as well that forgiveness has also emerged as an object of study across a range of different spheres, including psychoanalysis, psychology, and philosophy.[30] The emergence of Truth and Reconciliation commissions in places such as Rwanda, South Africa, and Canada have raised interest in the idea of the political and social values of forgiveness as necessary for moving on. In each of these cases the purposes are the same – to confront aspects of the past, to allow people to move on, to live without fear, to regain trust in other people, and to restore self-respect. In addition, each of these cases we can think about forgiveness as having to do with other people. Put a different way, it is other-oriented, to forgive someone is to think of someone else's situation. Of course, the other aspect of forgiveness is that it must be asked for.

For Hannah Arendt, action occurs in the "space of appearance." Also for Arendt, action is very closely tied to acts of speech. Forgiveness represents "the possible redemption from the prediction of irreversibility, of being unable to undo what one has done though one did not, and could not, have known what he was doing."[31] It is this notion of "irreversibility" that is important to Arendt's conception of forgiveness. Without forgiveness "our capacity to act would be confined to a single deed from which we could never recover," for she posits that it is the emergence of modern science and technology that "have carried irreversibility and human unpredictability into the natural realm, where no remedy can be found to undo what has been done before."[32] For Arendt the two human things that we can do are to keep promises and to forgive:

> Perhaps the most plausible argument that forgiving and acting are as closely connected as destroying and making comes from that aspect of forgiveness where the undoing of what was done seems to show the same revelatory character as the deed itself. Forgiving and the relationship it establishes is always an eminently personal (though not necessarily individual or private) affair in which what was done is forgiven for the sake of who did it.[33]

A key idea for Arendt is the absence of intimacy in her conception of forgiveness. It represents a form of respect from a distance; it occurs "without intimacy and without closeness: it is a regard for a person from the distance which the space of the world puts between us."[34]

An article on CNN's religion blog announced the release of "Confession: A Roman Catholic App." The application was created under the supervision of two Roman Catholic priests and it was intended to allow users the express contrition for their sins and to seek ways to atone for those sins. Passwords protect your identity and once you have confessed, your sins are apparently "wiped away." The Catholic Church, however, did not sanction the new app, saying that confessions must be made in person. "One cannot talk in any way about a 'confession via iPhone,'" a spokesman is quoted as saying. "This cannot in any way be substituted by a technology application."[35] However, an app for forgiveness is not that different from a much earlier Catholic practice of purchasing indulgences, which were used as an immediate and advance on confession before the face-to-face meeting that might occur in a church.

Could we also conceptualize things such as pardons as a kind of distant forgiveness?[36] In her account of "pardon tales" in sixteenth-century France, Natalie Zenon Davis describes such accounts as mixtures of judicial supplications, re-tellings of past actions, and, of course, narrative, all part of the recreation and recasting of past events. The pardon, once requested and granted, then offers a restorative effect, all through mediated communication:

> By a letter of remission (once it had been ratified by a court of law) the king's grace prevented a person from being so executed and also prevented or limited the royal confiscation of good that accompanied such a penalty, or – to quote the formula – "pardoned the act . . . and remitted every penalty, fine and corporal, criminal, and civil injury which might be incurred as a result of it . . . and restored to the supplicant his good name and reputation and his goods.[37]

Similar to pardons are acts that grant amnesty. Amnesty represents pardoning but to a wider audience in which some authoritative branch offers forgetting and forgiveness in return for some restorative act undertaken within a fixed period of time. Owners of semi-automatic weapons, for example, might escape prosecution if they return their guns to a police station during a particular period of time. For Peter Krapp, amnesties "allow one to go on 'as if nothing had happened,' imposing silence about the memory of the unforgettable."[38]

It is clear from this brief description that notions of pardon and amnesty are simply too onerous for the kinds of remedial work I have been discussing in this chapter. And yet the prospects of some form of reprieve that comes from a distance seems attractive for addressing the ethical issues associated with the distributed forms of communication and memory that exist in the present day. Seen this way, even in these basic terms, we can consider forgiveness as working against notions of perfect memory that the Internet may offer by recognizing embarrassments as natural consequences and representative of the Internet's imperfect memory. There may be more records of you doing things but there is no possible way that it will remember *everything* about us. What exists online can only ever be, at best, a partial view, only a part of the narrative of individual experience. Finally, such conceptions may lead us toward thinking about forgiveness without necessarily tying it to notions of forgetting. Avishai Margalit's *Ethics of Memory* explains that expecting forgetting to accompany forgiveness may not be reasonable; however, an emphasis on forgiveness *without* forgetting privileges an attempt to move not only beyond feelings of anger and vengefulness but also through the gift of forgiveness, to restore social bonds based on reconciliation.[39]

It may also represent a more profound moral stance upon which to make sense of living in a world where other people – strangers – know a lot about us. Roger Silverstone notes that we have become ambivalent and ambiguous about how to handle relations in online spaces because of a lack of knowing how to act in relation to each other. Some kind of distant forgiveness at least represents an attempt to tilt the balance away from those stances by offering the prospect of free passes for things that may be embarrassing.[40] Alison Winter recently suggested that somewhere between memory and forgetting lies the nineteenth-century notion of character. For Winter, character was a concept that viewed someone as being built up by daily experiences and the personal choices they made and in which character was "an accretion of memory."[41] You were, in a sense, a product of all of the things that you remember about yourself, even those things that may be embarrassing.

Clearly, what I have offered here is itself an imperfect set of solutions, but it represents an attempt at starting to develop a set of principles for addressing how we might be able to engage with each other against the backdrop of the expansiveness of digital memory. If "right to forget" policies represent highly imperfect and deeply problematic policy measures, they do represent an attempt to put the humans back into a conversation about living in a digital culture, and perhaps that begins by reclaiming character as being more than a unit of information. This alone represents a counterweight present in so many ways where we think about and appear to be dealing with what is being remembered about us online.

Notes

1 Mark Deuze, *Media Life* (Malden, Mass.: Polity Press, 2012).
2 Luke Purshouse, "Embarrassment: A Philosophical Analysis," *Philosophy* 76, no. 298 (October 2001): 515–40.
3 danah boyd, *It's Complicated: The Social Lives of Networked Teens* (New Haven: Yale University Press, 2014), 36.
4 Andrew Hoskins, "The End of Decay Time," *Memory Studies* 6, no. 4 (2013): 387.
5 Ori Schwarz, "The Past Next Door: Neighbourly Relations with Digital Memory Artifacts," *Memory Studies* 7, no.1 (2003): 7–21.
6 Roger Silverstone, *Media and Morality: On the Rise of the Mediapolis* (Malden, Mass.: Polity Press, 2007), 47.
7 Jose van Dijck, "Digital Photography: Communication, Identity, Memory," *Visual Communication* 7, no. 1 (2008): 57–76.
8 Alice Marwick, *Status Update: Celebrity, Publicity, and Branding in the Social Media Age* (N.H.: Yale University Press, 2013), 114.
9 Brian Stelter, "'Ashamed': Ex PR-Exec Justine Sacco Apologizes for AIDS in Africa Tweet," CNN.com, December 23, 2013, www.cnn.com/2013/12/22/world/sacco-offensive-tweet/.
10 Viktor Mayer-Schönberger, *Delete: The Virtue of Forgetting in the Digital Age* (Princeton, N.J.: Princeton University Press, 2009), 1.
11 Canadian Broadcasting Corporation, "Depressed Woman Loses Benefits over Facebook Photos," November 19, 2009, www.cbc.ca/news/canada/montreal/depressed-woman-loses-benefits-over-facebook-photos-1.861843.
12 Gabe Friedman, "The Biggest Filer of Copyright Lawsuits? This Erotica Web Site," *New Yorker.com*, May 15, 2014, www.newyorker.com/business/currency/the-biggest-filer-of-copyright-lawsuits-this-erotica-web-site.

13 John Durham Peters, "Mass Media," in *Critical Terms for Media Studies*, ed. W.J.T. Mitchell and Mark B.N. Hansen (Chicago: University of Chicago Press, 2010), 266–79.
14 Schwarz, "The Past Next Door," 10.
15 For an interesting discussion on stock photography, see Paul Frosh, *The Image Factory: Consumer Culture, Photography and the Visual Content Industry* (London: Bloomsbury Academic, 2003).
16 James Hutton, Michael Goodman, Jill Alexander, and Christina Genest, "Reputation Management: The New Face of Corporate Public Relations," *Public Relations Review* 27, no. 3 (2001): 247–248.
17 Reputation.com, "About Us," www.reputation.com/about-us.
18 "The New Reputation Custodians: Repositioning Individuals as the Guardians of Their Online Reputations." Paper presented to the Annual Conference of the International Communication Association, London, United Kingdom, May 2013.
19 AIM Sports Reputation Management, "Services," http://aimsportsrep.com/services/.
20 David Streitfeld, "European Court Lets Users Erase Records on Web," *New York Times*, May 13, 2014.
21 EU, Directorate General for Justice, *Charter of Fundamental Rights in the European Union*, http://ec.europa.eu/justice/fundamental-rights/charter/index_en.htm.
22 Silke Wünsch, "German Federal Court Raps Google on the Knuckles over Autocomplete Function," *Deutsche Welle*, May 15, 2013, www.dw.de/german-federal-court-raps-google-on-the-knuckles-over-autocomplete-function/a-16813363.
23 Mayer-Schönberger, *Delete*, 171.
24 van Dijck, "Digital Photography," 63.
25 van Dijck, "Digital Photography," 63.
26 All videos can be found at the "mini-site" for data-protection rules offered by the Justice Directorate of the European Commission, http://ec.europa.eu/justice/data-protection/minisite/users.html.
27 Viviane Reding, "The EU Data Protection Reform 2012: Making Europe the Standard Setter for Modern Data Protection Rules in the Digital Age" (January 22, 2012), http://europa.eu/rapid/pressReleasesAction.do?reference=SPEECH/12/26&format=PDF.
28 Benedict Anderson, *Imagined Communities* (London: Verso, 1991), 205.
29 Opinion of Advocate General, "Google Spain/Google Inc. v. Agencia Española de Protección de Datos (AEPD)/Mario Costeja González, June 25, 2013, http://curia.europa.eu/juris/documents.jsf?num=C-131/12.
30 Christel Fricke, "Introduction," in *The Ethics of Forgiveness: A Collection of Essays*, ed. Christel Fricke (London: Routledge, 2011), 2–3.
31 Hannah Arendt, *The Human Condition* (Chicago: University of Chicago Press, 1958), 213.
32 Arendt, *The Human Condition*, 214.
33 Arendt, *The Human Condition*, 217.
34 Arendt, *The Human Condition*, 235.
35 Hada Messia and Dan Gilgoff, "Vatican Issues Warning for New Confession App," *CNN.com* February 9, 2011, http://religion.blogs.cnn.com/2011/02/09/vatican-issues-warning-for-new-confession-app/.

36 I have adapted the notion of "distant forgiveness" from Luc Boltanski, *Distant Suffering: Morality, Media, and Politics* (Cambridge, Mass.: Cambridge University Press, 1999).
37 Natalie Zenon Davis, *Fiction in the Archives: Pardon Tales and Their Tellers in the Sixteenth-Century France* (Palo Alto, Calif.: Stanford University Press, 1987), 6
38 Peter Krapp, "Amnesty: Between an Ethics of Forgiveness and the Politics of Forgetting," *German Law Journal* 6, no. 1 (2005): 185–95.
39 Avishai Margalit, *Ethics of Memory* (Cambridge, Mass.: Harvard University Press, 2002).
40 Silverstone, *Media and Morality.*
41 Alison Winter, *Memory: Fragments of a Modern History* (Chicago: University of Chicago Press, 2012).

Chapter 9

Ambiguity, cinema and the digital documentary image

Roderick Coover

One of the values attributed to the arts is their articulation of ambiguity. Ambiguity unsettles the authority of exposition, rational argument and evidentiary media at levels both of form and content. Documentary arts that use mimetic technologies such photography, film and video have never escaped a paradox in that they may both present evidence and pose questions about the evidence they display; they can both constrain and articulate ambiguity. This paradox leads documentary to a unique position in serving fields across the sciences, humanities and fine arts. Cameras – mimetic, technological devices – capture more in a moment's moment than the human eye can register. Their evidence is constructed artificially through frames (or stills), and, in time-based media such as film, through clips and edited sequences. Cameras also introduce ambiguity. Photographic and cinematic images raise questions about the limits of sight, about what is in the frame and what seems to be beyond it, about what is objectively fact and what is subjectively attributed, and what simply can or cannot be filmed.

The frame, or still, which is a fundamental building block in discussions of how nineteenth- and twentieth-century mimetic devices work, ostensibly "captures" a moment; it is a shock that artificially arrests actions, objects, expressions etc. from the flow and flux of time. For cinema, another building block has been the clip, or take, and another has been the cut, montage or edit. Theories about documentary's evidentiary capacity are based on these concepts of the still, clip and montage. Drawing the stability of these concepts into question in light of technological developments of computing, this chapter reconsiders ambiguity in the documentary image.

The two images in Figure 9.1 picture boat launches of Major John Wesley Powell's exploration expeditions of the Colorado River. On May 24, 1869, a 35-year-old, one-armed, Civil War veteran, Major John Wesley Powell (1834–1902) and nine crewmembers began an exploring expedition of the Green and Colorado Rivers. They aimed at being the first known team to navigate and map the rivers and nearby canyonlands, some areas of which were sparsely populated by differing native peoples, pioneers, and, to the West, Mormon settlers. Rail service was opening the desert West to future

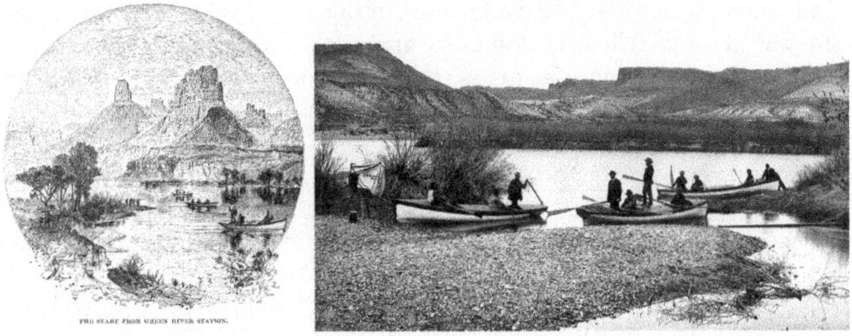

Figure 9.1 An etching by Thomas Moran and photograph by E.O. Beaman depicting John Wesley Powell and a crew departing near Green River City on an exploring expedition of the Green and Colorado Rivers.

settlement with the first transcontinental service meeting at Promontory Summit, Utah, only two weeks before Powell's launch, and the Major was hurrying to beat others to the opportunity to map and name the *terra incognita*.[1] The expedition launched from below a train trestle near Green River City, Wyoming. About 1,000 miles and 99 days later, the expedition concluded at the mouth of the Virgin River in southwestern Utah, although a few of the crew members decided to float onward down the calm and better-known waters of the lower Colorado toward Yuma and the Sea of Cortez.[2]

Thomas Moran's etching represents this historic 1869 departure in both official and popular documents.[3] However, there were no artists or photographers on the first trip, and the image is probably based on a photograph that was taken on a less-discussed second voyage that took place in 1871–1872. It is almost like a child's game to spot the differences between the illustration of the Wild West and the photo it came from. Bushes become trees, natural spires appear where none belong and a vast scenery is compressed into a single dramatic and romantic vision. One might wonder, what was wrong with the photo, which is also strange, beautiful and remote? Or, did it matter that the photo did *actually* document a second departure while the almost identical etching is timeless?

In 1873, Powell invited Thomas Moran to join a third expedition, this one a land expedition via the Virgin River and Kenab to the Grand Canyon. Moran came with his own funding from the journals *Appleton's*, *The Aldine* and *Scribner's Monthly*[4] and many of the etchings in Powell's publications carry his name. Decorating political institutions, parks and travel materials, Moran's images invoked classical themes of Western dominion.[5] Powell's works participated in this Caucasian-American romantic migration narrative and drew clear ties with Moran's sensibilities. While several crewmembers kept diaries of the first voyage, these were largely ignored in the narrative and report, which combined a memory of one trip with documentation from another. Powell relied on dominant approaches to writing and representation (with images, photos, maps etc.) that subsumed differing perspectives upon the *terra incognita* into relatively singular sensibility. Art and science blurred.

Fifteen years later, expectations of the use of imagery in official documents had changed. Topographic line drawings, maps and landscape photographs gained status as scientific evidence, and these supported geographic studies. Once a *terra incognita* to the migrating Caucasians, the great sandy plain was now established as land to be bought, owned, used, subdivided, reclaimed, irrigated, mined, developed or dammed and flooded. A West constructed in romantic imagery and narrative language was quickly fading into the stuff of history and legend – of popular media and collective imagination. New evidentiary representations were drawn in lines and grids and written out in deeds and laws – a very differently sensibility.[6]

The etching is iconic and symbolic and, like most such etchings, it prioritizes, augments and eliminates details to illustrate the primary fact or drama. Despite techniques such as framing, focus, light adjustments (and, in cinema, motion), which all create authority, the observational image inevitably includes more than it can control and offers conditional verisimilitude through that excess. Paradoxically, excess produces ambiguity.[7] Without clarification, one cannot know if the above photograph (Figure 9.1) was primarily made to record information about the individuals, the kinds of boats used on the journey (notably, an important error in this case, as the etching claims to depict the maiden voyage when actually a different kind of boat was used), the weather, the biology, the geology, the performance of departure or any other question. Identifying ambiguity in images engages the imagination in making meaning. Sometimes text shapes how a viewer looks at an image and what the viewer looks for. Sometimes other images offer context and direction, as is common in film, and sometimes ambiguities in the image are unravelled through other information, such as sound.[8]

A photographic and cinematic documentary image is not like sight. Humans do not "see it all" in a single flash the way the mimetic devices (cameras, for example) record time. Where the frame offers an excess of spatially organized information, the documentary clip generally offers temporal excess. There is generally more content in a motion image of actuality that a viewer can digest; the viewer, not unlike the maker, constructs understanding from the information flashing by. Film viewing mirrors, and perhaps articulates, a fundamental human process by which the mind distinguishes, sorts and connects sensory information in time to construct, and continually reconstruct, a more or less cohesive sense of whole. Documentary representations of actuality, such as those of most ethnography, are *constitutive*, as Dia Vaughan writes,[9] and if such information wasn't valued, they could be shot entirely in studios – with an effect analogous, perhaps, to reducing the documentary photography to an etching.

Where the frame establishes spatial relationships, the clip creates temporal ones. This, too, is a kind of authority. Film cameras introduce a time-constant – a point underlying Andre Bazin's praise of Flaherty's long takes of Nanook catching a seal in *Nanook of the North*. The action is all there to see in the frame and it all happens in time, which is to say that it happens in time as mediated by the constants of the camera and projector. The camera's supposed objectivity is its observation stance that imposes an omniscient view and cinematic time, drawing the action into the time-constant of the overall film.

Although, in cinema, frame rates may be altered for effect, in most instances of documentary production both in recording and in playback, frame rates are largely imperceptible constants. The single-channel, linear-time qualities of film and the sensation of regulated time contrast with other modes of media reception such as reading. Generally, single-channel projection media did not allow viewers to escape the temporal constraints – or authority – of the projection mechanism.[10] Actions may move at different rates with differing modal, tonal and rhythmic qualities in their takes and cuts; however, within one system or another the frame rate is usually constant. How was Nanook experiencing time while (supposedly) struggling with that supposed seal beneath the water? Illusions of verisimilitude are constructed in the clips through both continuity and contiguity, with consistent and inclusive framing re-affirming relationships in the *mise-en-scène* vis-à-vis the frame and the objective, distanced stance of the camera.

The documentary image is its excess, which is also recognized as shock. A captured image is shock against the flows experienced by consciousness.[11] This excess and shock provided by technology are described in various ways such as through Benjamin's concept of the optical unconscious,[12] and the shock is articulated well in the haunting magic of the freeze-frame, such as in Vertov's film, *Man with a Movie Camera* (Figure 9.2). The cut, or edit, in documentary film (or photographic sequences, too) is paradoxical. The cut offers clarity to ambiguities by providing connections and allowing viewers to construct meaning.[13] However, the cut in single-channel film is also the artifice that disrupts contiguity and its verisimilitude. The cut offers only one trajectory at the loss of others and, therefore, it dictates its own predications, leading viewers to draw a connection to what seems to be coming ahead from what has just past. Thus, the cut creates absence and loss; the potential of the image – its ambiguities – is developed in one direction at the expense of others. In documentary films and videos, the cut supports many goals including constructing stories and illuminating details (as in films such as *Nanook of the North* and *Grass*), suggesting symbolic or tropic connections (as in *Song of Ceylon* and *Les Maitres Fous*), expressing patterns and movements (as in *Man with a Movie Camera* and *Berlin: Symphony of a Great City*) and

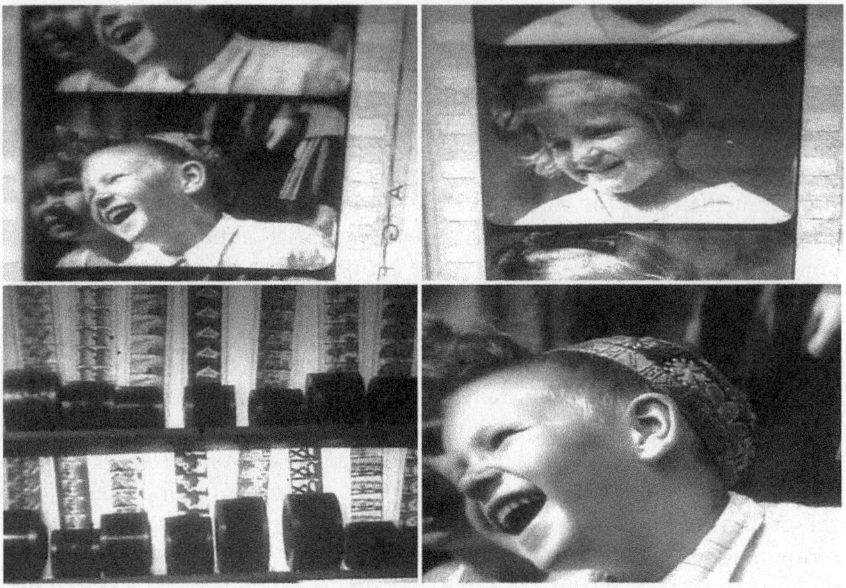

Figure 9.2 Still from Dziga Vertov's *Man with a Movie Camera*, in a sequence that demonstrates how film is made of frames that are cut, edited and animated.

offering argument or exposition (as in *The Plow that Broke the Plains* and *Sans Soleil*) etc.

Although many uses of digital technologies in documentary arts copy and look like prior forms, there are differences that challenge fundamental premises of what constitutes a cinematic image. The building blocks of film theory – the frame, clip and cut – prove to be malleable or even absent. These changes challenge over a century of film theory built upon these concepts including concepts of evidence, subjectivity and ambiguity in various forms of documentary representation. This topic is explored in a series of digital works I created about exploration, landscapes and their representation entitled *Unknown Territories*. The first of these works, *Voyage into the Unknown* (Figure 9.3) concerns John Wesley Powell's exploring expedition and the representation of that expedition through conventions of writing, illustration and photography. Later sections also concern uses of fiction, video, maps and brochures.

The digital format offers some valuable alternatives to print and single-channel video, enhancing qualities of ambiguity through the inclusion of differing perspectives, materials, disciplinary modes of inquiry and/or paths. One example of how a digital format can impact on representation is the interactive work *Unknown Territories*[14] in which documentary materials about the representation and conceptualization of the arid West are gathered in narrativized scrolling environments. *Unknown Territories* explores how perceptions of place are shaped diversely through differing modes in writing and the arts; these can be engaged in digital environments in ways that differ from their engagement in other formats. In looking at evidence and imagery from John Wesley Powell's exploring expeditions as well as later materials, the project asks how does one come to know and imagine an "unknown territory"?[15]

Paths cut across history and landscapes as reader-viewers take up questions of growth and migration, industry and mining, including the impact of the uranium boom, dam building and tourism, all explored from the perspective of the humanities, to ask how are these varying historical developments and arguments framed through language and image. For example, in *Voyage into the Unknown* users navigate an interactive environment as they travel with the one-armed Major Powell and his crew of Civil War veterans down the then-uncharted waters of the mighty Colorado River into a *terra incognita*. First comes the adventure, then comes its representation. Much later comes critical examination and, perhaps, as a whole, re-invention. The work offers a learning environment that integrates readerly and viewing experiences. Part narrative, part documentary, the work bridges modes of writing and image-making through the use of a sequentially loading landscape and Adobe Flash® movie-based segments. Through interactive features, users discover how events and diaristic observations later become recast through photographs, illustrations, articles and books. The work draws its users' attentions to how differing media that were used by Powell and his colleagues might have contributed to popular conceptualizations of the American landscape.

The landscapes of *Unknown Territories* literalize the metaphor of making paths. In these panoramic environments, time-based cognition and text-based knowledge acquisition may go hand in hand, or may clash like objects in a collage, colliding in a field on which, at first glance, they would not seem to belong. The user takes a journey along possible routes and departures. The user is gathering information en route and piecing together a narrative. What do these elements add up to? These works confronts the very question revealed by the many ways evidence of the original journey is transformed after the fact; there can be no simple, monological or monolithic summation. Choice-making is fully integrated in the user experience allowing individuals flexibility in choosing to expand or limit narrative, expository and poetic approaches to a documentary's primary topic and its off-shoots. This off-shoot leads viewers into specific topic segments with narratives of their own.

The borders between reading and viewing blur in an active process that reveals its construction and expository processes, leading viewers to engage with the director in the process of constructing meaning out of experience. On the one hand, propositions and arguments are developed through montages that provide a route through the material – a director's cut. On the other

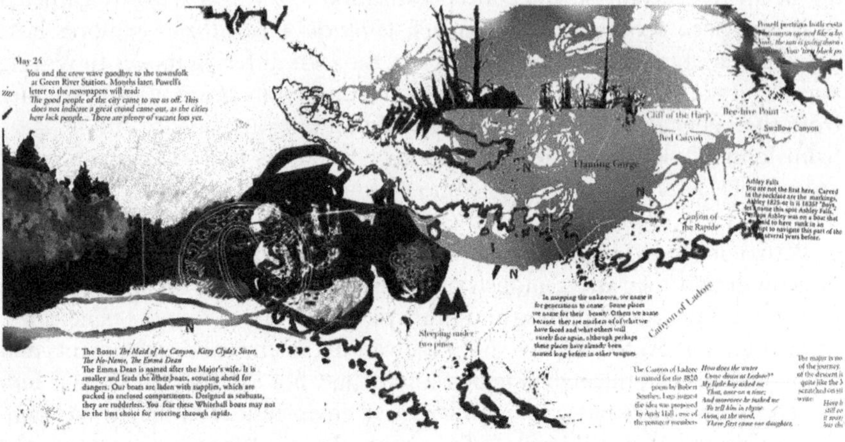

Figure 9.3 Sample from Roderick Coover's *Voyage into the Unknown*.[16]

hand, unlike a linear, single-channel work, viewers may alternatively navigate through clips, interviews and other materials through parallel or diverging paths to arrive at differing and multidisciplinary perspectives on shared questions of action, experience and knowledge. Composed of layered tropes, juxtaposed paths, modally varied arguments and active choice-making, the emerging rhetoric and poetics of documentary in the new media have been scouted by some, but they still remain, for the most part, uncharted . . . a *terra incognita*.

These changing ways of making sense of differing media are constructed through layered tropes, juxtaposed paths, modally varied arguments and active choice-making, both in production, by the makers of media works, and by users, who may in turn adopt elements even directly within a work. This trend continues with social media and adaptable mobile forms. Clips – often unedited ones – proliferate in evidentiary exchanges and contexts, and the narratives binding them are often organically and socially created. Hybrid spaces that combine text and video in shared environments challenge single-channel cinematic conventions of linearity and montage, and database, generative or constitutive films undermine the monological imperatives common in cinema. The dichotomy of the clip and the cut that characterized but was never resolved in many of the great texts of film theory – such as Eisenstein[17] and Bazin[18] – dissolves when, in digital works, long takes coexist with montage sequences (and with lots of other information as well).

The practical result for the creators of motion media works is that much of what used to end up on the cutting-room floor or as little-used master video tapes on the edit room bookshelves is now available. In interactive works, first, users may have the opportunity to access source materials and judge the maker's choice-making. Second, interactive formats enable video material to be combined with text elements such as original writing, field notes, primary documents, secondary documents, interview transcripts and so forth. Third, digital tools facilitate the inclusion of other kinds of visual materials into a project, such as maps and photographs, which can be compared with cinematic representations. Practically speaking, questions of design are not so different from those of writing or editing; the ethnographer/media-maker creates paths (arguments) via research materials. One difference for interactive works, however, is that the research materials may be included and the processes of research and representation may be revealed. Users may be able to follow the maker's choices and decisions that went into building an argument out of the fragments of experience and data. In the face of alternatives, the researcher encounters an increased need to present supporting evidence for how and why particular routes through the material were valuable, but may face less pressure to gel materials in narrative or expository constructs.[19]

In *Another Way of Telling*, Berger writes that an important difference between viewing (or reading) images in a book and watching such images in a film, is the forward temporal force of the technology, which Berger characterizes as producing a kind of temporal anxiety caused by the technological provocation to attend to each forthcoming frame. Berger writes:

> Eisenstein once spoke of a "montage of attractions". By this he meant that what precedes the film-cut should attract what follows it, and vice versa. The energy of this attraction could take the form of a contrast, an equivalence, a conflict, a recurrence. In each case, the cut becomes eloquent and functions like the hinge of a metaphor. . . . Yet there was in fact an intrinsic difficulty in applying this idea to film. In a film there is always a third energy in play: that of the reel, that of the film's running through time. And so the two attractions in a film montage are never equal. . . . In a sequence of still photographs, however, the energy of attraction, either side of a cut, does remain equal, two way and mutual. . . . The sequence has become a field of coexistence like the field of memory. . . . Photographs so placed are restored to a living context: not of course to the original context from which they were taken – that is impossible – but to a context of experience.[20]

Navigation allows users to cross-reference images to discover formal, tropic, narrative and expository significations. The ability to juxtapose and link diverse kinds of materials expands the potential for reflexivity. The navigable spatial arrangement of the book enables choice and subjective temporality. Ambiguity in the image is restored.

Perhaps film editing is, and has always been, hypertextual. In celluloid editing practices, clips are examined as discreet physical objects that hang from bins or are coiled on cores. They are arranged and often re-arranged into sets, which are spatial configurations, and they are given tags and annotations through logs.[21] The clips are gathered, taped and later glued in various physical variations. The editor fingers and scrolls through these, at times making cuts as much by the physical lengths of the clips as by their contents. Likewise, digital editing environments also arrange clips, or more correctly icons that signify clips, spatially. Bins, timelines and menus represent forms of spatial organization from which temporal experiences of actually watching clips are triggered. Where time-based viewership largely stimulates spontaneous constitutive processes (or, "worldmaking"), editing and other hypermedia activities more significantly emphasize conscious and reflexive constitutive processes in which questions that are raised by one image get explored through another, or another, or another. The editing process requires choice-making and selection. The editor may imagine and create sequences from clips in almost infinite variations, even if, in the final result, all but one of those variations are discarded, and the rejects are forgotten along with the myriad lessons and alternatives they may have offered. Digital technologies enable the inclusion of materials recorded or organized through differing modes as well as the incorporation of other kinds of research materials such as text, maps and photographs. They can allow for continual updating and offer opportunities for using algorithms to create versions generated by the computer or user inputs. Further, in locative media projects, virtual "edits" may even be created by users physically walking among actual places, conjoining located materials en route. Therefore, the editor may also be a theoretician, technician, writer, explorer, researcher and designer, and this may result in projects that are equally experiential or intellectual.[22]

As the format of interactive editing programs, interactive documentaries, web-based clips and database films merge with those of writing, data collection and spatial presentation, a new temporal relationship to argument emerges. Bringing together differing kinds of research materials can enable users to follow the media maker's process, whether by allowing users to read field notes and supporting documents or to follow how particular sets of materials led to the development of an edit or argument.[23] When supporting materials and data are available, the user can see how choices were made and consider alternatives.[24] The media maker is not deprived of the power to make an argument and have a voice (expressing one's ideas is among the important reasons that individuals make works). In fact, the maker may offer many arguments that would not fit together in the logics of a single-channel work. As evidenced in works by the Labyrinth Project, Jeffrey Shaw, Flavia Caviezel, Kurt Fendt and Ellen Crocker[25] among many others, such works can express relationships between the user, maker and subject that raise interesting ethical questions about single-channel media and the messages they may convey through form. Furthermore, there are many ways of working with images, gathering visual evidence and layering them made possible by mobile technology and other kinds of organizational frameworks, such as Google maps for example.

One might conclude that new concepts in digital video are contributing to a recognition of ambiguity in the structures of mimetic representation. New media can be used to suggest differing temporal and modal characteristics of visual experience. They may liberate elements from the regulation of the frame and frame rate. Digital formats change user and maker relationships to imagery.[26] In some works, the film editor may explore new ambiguities in the editing process, and the viewer may now have opportunities to do so, too.

However, paradoxically, some provocations of ambiguity unique to cinematic film may seem diminished. Take, for example the particular experience created through the flicker of film played at 18–24fps (frames per second). There is also a perceptibly different feel between 24fps projected film and emergent 48p digital formats. The flicker of old film is itself a kind of temporal ambiguity pointing to a break between frames that is both absent from the image and perceptible in its absence. Equally illusive are the qualities in shadows that also introduce ambiguity into images. As good as digital video forms are, they do not express shadows in quite the same way. This is particularly noticeable in comparing digital images to projected black and white cinematic ones. The lack of color specificity and a richness in shadows can provide a high capacity for abstraction and tropic predication. Therefore, the articulation of some kinds of ambiguity may be achieved through new media and new techniques such as those of layering, compositing and interactivity but others may be concealed.

Notes

1 Powell first visited the area two years earlier in 1867, when, as a professor of natural sciences at Illinois State Normal University, he led a group amateur geologists and students to collect rock samples in Colorado. He set his ambitions on returning to explore the waterways of the Colorado River, which he began mostly with limited funding from sources in Illinois and his own salary. For more see William deBuys, "Introduction: Seeing Things Whole," in *Seeing Things Whole: The Essential John Wesley Powell*, ed. William deBuys (Washington, D.C.: Island Press, 2001), 1–24.

2 For an expanded discussion of this voyage, see Roderick Coover, "Picturing the Great Unknown: John Wesley Powell and the Divergent Paths of Art and Science in the Representation of the Colorado River and Utah Canyonlands," in *Writing with Photography*, ed. Liliane Weissberg and Karen Beckman (Minn. University of Minnesota, 2013), Chapter 2.

3 See, for example, John Wesley Powell, *Exploration of the Colorado River of the West and Its Tributaries* (Washington, DC: US Government Printing Office, 1875) and John Wesley Powell, *Canyons of the Colorado* (Meadville, Pa.: Flood & Vincent, 1895).

4 The circumstances described in letters by Thomas Moran in *Home-thoughts, from Afar: Letters of Thomas Moran to Mary Nimmo Moran*, ed. Amy Bassford and Fritiof Fryxell (Davenport, Iowa: Wagner's Printers, 1967), 41–42. These are further discussed in Elizabeth C. Childs, "Time's Profile: John Wesley Powell, Art, and Geology at the Grand Canyon," *American Art* 10, no. 1 (1996): 7–35.

5 Curiously, Moran's painting of the Grand Canyon is less successful than his painting of Yellowstone because it does not fill the proper myth of the West. It is critiqued in the *New York Times* and *Atlantic Monthly* for expressing the West as a barren wasteland – a place of nightmares not dreams, in William H. Truettner, "Scenes of Majesty and Enduring Interest: Thomas Moran Goes West," *The Art Bulletin* 58, no. 2 (1976): 241–259.

6 For more, see Roderick Coover, "Picturing the Great Unknown."

7 For more about ambiguity in images, see John Berger and Jean Mohr, *Another Way of Telling* (New York: Pantheon, 1982). Meanings of a lone photograph, they argue, are empty and are contextualized by knowledge, text or other photographs, by which one may begin to identify correspondences, articulate ambiguities and make meaning. Curiously, the fact that Berger and Mohr make their argument through book-printed photographic images suggests there is some degree of commonality across printing forms. The relationship of differing photographic print and projection forms on visual messages is a valuable topic for another paper.

8 These ambiguities are expanded in the case of 1870s geographic photography, as the photographs, ostensibly as evidence of landscape, often included individuals and exploration in participation with constructions of cultural histories and myth-making of US Western exploration and migration.

9 If images didn't offer more than what could be imagined and controlled, Vaughan points out, ethnographies could be entirely made in studios with actors. For more on the constitutive nature of documentary reception, see Dai Vaughan, *For Documentary* (Los Angeles: University of California Press, 1999).

10 For more, see Roderick Coover, "Interactive Media Representation," in *The Sage Handbook of Visual Research Methods*, ed. Eric Margolis and Luc Pauwels (London: Sage Publications, 2011), 619–637.

11 Humans to do not have photographic memory, and they do not see all things equally. Nelson Goodman defines processes of making sense of stimuli as "worldmaking" in Nelson Goodman, *Ways of Worldmaking* (Indianapolis: Hackett Publishing Company, 1978). According to Goodman, seeing is motivated, and it is built upon memory. The human mind chooses what to look at and what to ignore, making sense of information through at least five processes: (1) composition and deletion; (2) weighting; (3) ordering; (4) deletion and supplementation; and (5) deformation. Through these processes, new realities are built out of old ones. For more on Nelson Goodman's theories of "worldmaking" and their relation to cinematic practices of editing and anthropological theories of tropes, see Roderick Coover, "Worldmaking, Metaphors, and Montage in the Representation of Cultures: Cross-Cultural Filmmaking and the Poetics of Robert Gardner's Forest of Bliss," *Visual Anthropology* 14, no. 4 (2001): 415–438.

12 Walter Benjamin, "A Short History of Photography," in *Classic Essays on Photography*, ed. Alan Trachtenberg (New Haven: Leete Island Books, 1980), 199–217.

13 For more on the cinematic construction of meaning through tropic predication see Roderick Coover, *Cultures in Webs: Working in Hypermedia with the Documentary Image*, Interactive CD-ROM (Watertown, Mass.: Eastgate, 2003).
14 Roderick Coover, *Unknown Territories* (2012), http://unknownterritories.com.
15 For further discussion see Roderick Coover, "Taking a Scroll: Text, Image and the Construction of Meaning in a Digital Panorama," *Hyperrhiz: New Media Cultures* no. 6 (2009), www.hyperrhiz.net/hyperrhiz06, which directly expands some of the discussion in this section and Coover, "Interactive Media Representation," which considers these theoretical questions in relation to other visual research projects.
16 Roderick Coover, *Voyage into the Unknown* (2007), http://unknownterritories.com.
17 See, for example, Sergei Eisenstein, *The Film Sense* (New York: Harcourt, Brace and World, 1975) and Sergei Eisenstein, *Film Form: Essays in Film Theory* (San Diego: Harcourt, 1977).
18 See, for example, Andre Bazin, *What Is Cinema?, Vol. 1* (Los Angeles: University of California Press, 1967).
19 For more, see Coover, "Interactive Media Representation."
20 Berger and Mohr, *Another Way of Telling*, 288–289.
21 As a result of the conditions of videotape, analog video editing makes greater use of logs that are referenced to stacks of time-coded tapes.
22 This by no means rejects the importance of film, video, photography or other media; there remain expressions far better articulated in an optically printed film or a silver nitrate photograph, for example, than through a digital work. Each stimulates differing perceptive and cognitive responses and each expands the world in differing ways, articulating and engaging life's experiential ambiguities.
23 For more, see Coover, *Cultures in Webs*.
24 For more, see Roderick Coover, "The Digital Panorama and Cinemascapes," *Switching Codes: Thinking through Digital Technology in the Humanities and Arts*, in ed. Thomas Bartscherer and Roderick Coover (Chicago: University of Chicago Press, 2011), 199–217.
25 See Jeffrey Shaw et al., "Re-place: The Embodiment of Virtual Space," in *Switching Codes*, 218–237.
26 Roderick Coover et al., "Digital Technologies, Visual Research and the Nonfiction Image," in *Advances in Visual Methodology*, ed. Sarah Pink (Los Angeles: Sage, 2012), 191–208.

Part III

Embodying images

Chapter 10

Introduction

Subjectification as embodiment; subjectification is embodiment

Alexandra Sastre and Nicholas Gilewicz

Over the past decade, Charles Taber and Milton Lodge have examined the hot cognition hypothesis,[1] concluding that *affect* – determined by a range of factors from political agreeability to agreeable appearance – manifests itself in political decision-making processes from initial exposure to information all the way through claims to make rational choices when voting. Their findings should discomfit those wedded to the idea of politically rational voters, but propagandists, semioticians, and cultural studies scholars should find them unsurprising. Those fields have long emphasized the power of images – both rhetorical power and political power. For affect itself should be considered bodily in nature. If, as Taber and Lodge suggest, the affective is precognitive, perhaps it is best thought of as both psychological *and* physiological.

The power of images lies precisely in their ability to produce subjects and to subjectify: to effect our thoughts and affect our bodies. The four authors in this part make this power quite clear, as each highlights in his or her own way the oft-elided *processes* of subjectification that can govern the creations of images and our understandings of perception. By calling attention – sometimes obliquely, sometimes directly – to *how* those images are produced, the contributions of this part lay the groundwork for political and biopolitical considerations of the role of the image in subjectification.

Further, how the subject is perceived is how the subject is produced. In this way, the need to consider ethics of image production and consumption – as well as the image itself, as the node of power flows – becomes increasingly important. Images take people as their subjects, and that subjectification informs how we see the bodies and perceive the minds of those people, remaining real despite the image.

Orit Halpern's chapter tackles the 1959 installation "Glimpses of the U.S.A.," designed by Charles and Ray Eames for the United States Information Agency-planned cultural exchange with the Soviet Union; she investigates questions of perception and cognition through the role of cybernetics and then-new thinking about how images are processed and presented. Halpern considers how "algorithmic," "autonomous," and "potentially inhuman" notions of visual perception provided the logic of "Glimpses," a seven-screen installation (accompanied by narration, hanging dramatically from a Buckminster Fuller-designed geodesic dome over the equally famous Edward Steichen-curated "Family of Man" exhibit of still photography). Tracing the installation's rhythmic presentation of slices of American life, from aerial photographs to the winking eye of Marilyn Monroe, Halpern suggests that the Eamses were influenced by a study showing that the eye of the frog is capable of some amount of cognition independent of the brain – of identifying, as she writes, prey or enemies.[2] "Glimpses," in turn, was designed to capture the attention of our own roving eye, perhaps separately from our neural cognition. "For the Eameses," Halpern writes, "there was only communication: information, not representation." So what, then, are the implications of design predicated on a kind of autonomous vision?

A humanist response might be to argue that the separation of the eye from the brain is a separation, in fact, of sight from insight. This separation permits the integration of avant-garde techniques into a propaganda project, of which "Glimpses" was but one part. If decoupling sight from insight undergirds this installation, then the screens and even the cameras used to produce the films are arguably decoupled from critical and dialogical minds – a cleaving well-attuned to the propagandistic role of "Glimpses." Let no one claim the Eameses' designs were ever ill-considered.

The changing landscape of Kansas is the unexpectedly moving and provocative center of Lisa Cartwright and Steven Rubin's chapter "Sensory Topographies of Wind and Power in Kansas." Troubling a narrative of the changing heartland that merely replaces traditional farming life with the advancements of wind technology, she uses the ethnographic image to illuminate this palimpsest, where old and new find tenuous harmony in ways often invisible to the untrained eye. Engaging with the familial, gendered, and technological dimensions of Kansas's changing labor economy, Cartwright uses the camera as a pivotal tool in her "sensory ethnography," allowing the images to render palpable what she terms the "affective genotext" of this unique environment. The camera captures moments of camaraderie between instructors in this new technology and their local students, revealing new intimacies between man and machine. This revelation, in turn, complicates a particular legacy of technological advancement and opens up the possibility for reading symbiosis, beauty, and affect into this relationality. By capturing the minutiae of the local, Cartwright and Rubin's photographs concurrently recast simplistic narratives about American agrarianism and reflect global flows of power. Like the wind, the metaphoric and literal force that flows through her project, this matrix of power is integral yet invisible, and the still image, as a critical dimension of ethnographic work, reflects ongoing efforts to capture and engage its dynamism.

Amit Pinchevski's exploration of the face as the fundamental medium of communication asserts the exponential power of the mediated image in our contemporary world. "Never before," he writes, "have people seen so many of their own species as we and our most recent ancestors have since the advent of photography, film, and television." Modern imaging technology and modes of global connectivity have rendered the face imminently visible and facilitated its circulation through myriad spaces. However, today, the digitized visage has also become pure data, manipulable by algorithmic means. Consequently, Pinchevski argues, a paradox emerges: the very space–time collapse that fosters intimacy by giving us access to faces around the world we might otherwise never see also dematerializes and renders vulnerable their humanity.

In this way, not simply the image but the face itself is subjected to more flows of power than ever before. Turning to Hannah Arendt's notion of the "space of appearance,"[3] Pinchevski reminds us that being seen is inherently political. Critically, it is in these circuits of visibility that subjectivity is made, unmade, and remade, increasingly by agents largely invisible to the naked eye. For Pinchevski, the face is thus a vital heuristic for understanding the ethical stakes of our current media landscape. It is a call to remember that media do not only facilitate increased communication between people, but also shape the very parameters and *subjects* of those exchanges.

The image-as-medium has a similar dual nature, connecting and standing between. But it *also* produces, in the way that communication may be considered to constitute culture and society themselves, if we follow James Carey's observation, elaborating Raymond Williams: "If one tries to examine society as a form of communication, one sees it as a process whereby reality is created, shared, modified, and preserved."[4] A medium, considered through another definition, and echoing the mutually constitutive nature of communication and society, is also a place for growth – it is itself a space of production, and the image is no different. The image is a medium where communication may flourish, but at the same time it is a place where communication may be cultivated.

That is to say, while media are indeed growth agents, cultivation also implies personal agents, who exchange and produce (and worry over) our *understandings* of each other, through depictions. Consider Nazi obsessions with so-called "degenerate art," punctuated by the eponymous 1937 Munich exhibition. Consider the early-2010s fascination with the "thigh gap" and pro-anorexia websites. Consider the presentation of photographs of black Americans charged with crimes, perhaps most famously how the US newsmagazine *Time* darkened the tone of a police photo of the former American football player O.J. Simpson, accused of murdering his wife in 1994. Many more examples come to mind, but the point should be clear: images connect and stand between, but they also produce. They reflect and represent bodies, by embodying our ideas of each other. The process is iterative and interpolative, and demands that we consider the ethics of both product and production.

Together, these chapters prompt readers to think critically about not only how an image tells a story, but also who is doing the writing. They show us that images can trouble the things we feel we already know, but also that they can capture potential as yet unimagined. As Cartwright contends when chronicling a lesson in the new, vendible, and dangerous practice of "aerial search and rescue" emerging from the economy of wind, the image is a conduit for grappling with elements as seemingly disparate as gender, labor, history, and topography. Through the image, we are able to see both how intricate, global systems conjure new (and perhaps precarious) futures, and how they "literally touch the ground." And, through these studies, we are able to see that it is ever more pressing to recognize that the ethics of images have bodily stakes.

Notes

1 James P. Morris, Nancy K. Squires, Charles S. Taber, and Milton Lodge, "Activiation of Political Attitudes: a Psychophysiological Examination of the Hot Cognition Hypothesis," *Political Psychology* 24, no. 4 (2003): 727–745. Milton Lodge and Charles S. Taber, "The Automaticity of Affect for Political Leaders, Groups, and Issues: an Experimental Test of the Hot Cognition Hypothesis," *Political Psychology* 26, no. 3 (2005): 455–482. Milton Lodge and Charles S. Taber, *The Rationalizing Voter* (Cambridge: Cambridge University Press, 2013).

2 J.Y. Lettvin, H.R. Maturana, W.S. McCulloch, and W.H. Pitts, "What the Frog's Eye Tells the Frog's Brain." *Proceedings of the IRE* 47 (1959): 1940–1951.

3 Hannah Arendt, *The Human Condition* (Chicago: University of Chicago Press, 1958).

4 James W. Carey, *Communication as Culture, Revised Edition: Essays on Media and Society* (New York: Routledge, 2009), 26.

Chapter 11

The autonomy of the eye

Neuro-politics and population in design and cybernetics

Orit Halpern

My proposal in this chapter is to address the question of what is historically specific to our contemporary forms of image making, perception, and cognition, and to further ask how our history of these practices informs our future imaginaries, or what I would label "ethics" of the image. It is my contention that post-war design and computational sciences, particularly cybernetics, reconfigured the idea and practices of visuality and cognition in a manner that made perception itself, a medium – material, abstractable, and technically replicable – in unique ways that separated from earlier histories of subjectivity and objectivity.

In this chapter, I will address this challenge by examining one well-studied case of design informed through cybernetic and communication sciences – the 1959 installation "Glimpses of the U.S.A." The piece, built by the United States Information Agency (USIA) for a cultural exchange with the Soviets in Moscow, offers an extended exploration into how vision, image, and empire are closely entangled. As an archeological site, it also offers a historically specific meditation on the disjuncture between vision and human beings. This installation has been often discussed, most particularly, in the excellent work of Beatriz Colomina, as a site demonstrating new forms of media architectures and spectatorship grounded in information and communication sciences and enacting spectacles of data inundation.[1] I want to extend and complicate her account to include and engage the nascent cognitive and neuro-sciences of the period, race, and biopolitics. I do so to argue that a close inspection of this moment in design clarifies the discontinuities between the spectacles of the late twentieth-century and earlier cinematic and photographic apparatuses, while making visible the specific logics and practices that underpin what should be regarded as a form of autonomous sensory perception. In fact, I argue this emergent media infrastructure should not even be thought of solely in terms of anthropocentrism or the subjectivity. It is not the insufficiencies or lacks in human-embodied attention, as so regularly posited by disciplinary theories of control, that concerned designers, but rather the reconfiguration of the nervous system, from cognition to sensation, as the very medium, and material, for design.

As we shall see, the designers, and scientific theories, who built this piece insisted on the inhuman nature of visual perception. Simultaneously, however, the pavilion as a whole contained archival elements of documentary work, and aesthetics, that clarify contemporary forms of vision as temporally layered; this layering is not merely an accessory, but a fundamental aspect of producing images and screens. These two imperatives – to archive and to circulate – substantiate our contemporary turn toward seeking patterns in data and "visualizing" information. These productive tensions emerge around a new mandate at the time to produce an interactive mode of viewing.

Not surprisingly, biology, race, and gender play critical roles in conditioning spectatorship within this architecture. Designers worked with models of vision based in cybernetics and communication sciences, and sought to use their aesthetic tactics to present a racially diverse, and integrated, United States. In merging the representation of race with the science of machines, the final effect is to insist not only on the limits of human vision, but also to produce new ideas of species and territories, linked through literally nervous stimulation and speculation – a new neuro-political situation – that goes beyond the initial bio-politic formulation of subject and population. Population, here, is not the target, but rather the constituent, of media.

However, the recombination of archival vestiges of identity and representation and the introduction of communication design principles also results in a series of rarely explored, but none the less possible, multiplications in the image that destabilize assumptions about the direct relationship between computational abstraction and violence. If, as Jacques Rancière argues, politics is the organization of the sensible, then these multiple forms of organizing images translate into different political possibilities and prompt ethical questions about contemporary media practice, and how different interfaces effect and refract relationships between subjects and between populations within highly interactive environments.[2]

Political spectacle

In 1959, in Sokolniki Park in Moscow, the United States and the Soviet Union embarked upon the first of a series of programs in "cultural exchange." The Soviet Union would shortly thereafter have a pavilion in Columbus Circle in New York City. The program followed a series of global aesthetic confrontations staged at World Fairs, and the super powers were intent on producing as many forms of novel demonstration and presentation as possible.

For the American pavilion, a range of spectatorial experiences were prepared. Five hundred American corporations displayed their wares, along with models of the "splitnik," the Levitt-style suburban tract homes into which these commodities were to be placed. Fashion shows paraded the ideal nuclear American family, selected for its lack of exceptional talent, and, might I add, thinness. Beneath the images of commodities and gigantic screens of American landscapes was the pioneering photo-essay "The Family of Man" documenting a biologically diverse but still singular and universal human species.[3] Khrushchev and Nixon debated the variable merits of American and Soviet kitchen design and technology in front of the new television cameras.[4]

The highlight of the exhibition, however, was a huge multi-screen channel exhibit. The installation comprised of a vast cavern, built by Buckminster Fuller, filled with seven screens showing a multi-media piece, "Glimpses of the U.S.A.," designed by Charles and Ray Eames. A "totally new type of presentation," in the words of its designers, it was envisioned as a "letter," perhaps of love, between two nations in a world where writing would no longer suffice: a "glimpse" of the USA, a day in this foreign country's "life" that, by the end, would cease to be foreign. In the face of this imagined textual and lingual collapse, the designers believed visual images might serve as a new mode of human interaction. With design principles and aesthetics characterized by faith in the virtues of information inundation, the Eameses believed in spectatorship as choice. The idea of spectatorship invoked in their work was to force users to interactively pick a pattern through a vast data field. Some 2,200 images were shown on seven screens for 13 minutes. Charles Eames, in discussing his design and pedagogy for engineering, architecture, and business students, argued that vision was a new "language" and that the function of his multi-media displays was to test "how much information *could* be given" to a spectator in an allotted time.[5] Hundreds of thousands, if not millions, of people saw the installation.[6] Everyone, according to the art historian Peter Blake, "had tears in their eyes as they came out" of the opening show, including, rumor has it, Khrushchev. American officials were rumored to have called it one of the most successful acts of psychological warfare ever conducted. Ray Eames called it an "affective" experience.[7]

Despite the universal acclaim for the success of the American pavilion, it was deeply contentious, emerging from a series of accidents and assembled interests. While officials at USIA claimed that the exhibition was about promoting "improved understanding" between the two nations, this was hardly the only agenda. William Benton, a former Assistant Secretary of State, remarked that the State Department, "was in the propaganda not art business."[8] The art exhibition provoked attacks from a US Congress worried about whether artists were Communists.

More broadly, the United States struggled with its "soft" tactics. On the one hand, American government officials realized the salience and popularity of American products and entertainments; on the other, the desire to educate against the evils of Communism warred with the pleasures of film entertainments that often offered a vision of the United States as one of fast women and criminal men, obsessed by material wealth and bereft of soul.[9] No matter, the show would go on.

Scales of the image

The installation opens with the view from an aerial flight. "In the night sky," the voice over intones, "Russian and American cities look the same." The scene from the plane is ominous in the midst of debates over thermo-nuclear war. But this view from above rapidly condenses into the network. The history of planes over cities serves only as an aesthetic device for entering a circuit of images, not for ruminating on a history of war. Central are images of highways, modular housing, speeding cars and transport, infrastructures of power and industrial plants, conspicuous consumption, seven screens of Marilyn Monroe winking from *Some Like It Hot* (the Russians clapped every time), and, perhaps most critically, intimations of the perfectly racially integrated society. All cycling to a final moment where the screens turn black, and one image on only one screen appears – a single purple flower illuminated that final screen: the final signal. It was a "forget-me-not" flower; chosen for its symbolic connotations of love, friendship, and remembrance. In Russian, the name of the flower – *nezabutki* – translates literally as "forget-me-not." All questions of language and translation erased. The irony: the very infrastructure for an emerging American spatial apartheid, highways, and modular tract housing was the substrate for a collective vision of a single humanity. But the movie's major statement was that flow, and communication, will overcome difference – between nations, between people.[10]

Leading up to this enormous installation lay another path-breaking moment in the history of visuality – "The Family of Man." Curated by Edward Steichen, the chief photography curator of Museum of Modern Art, based on the 1930s documentary aesthetics of government-sponsored photography projects of the time such as those by the Works Progress Administration and the Farm Security Administration (in which Steichen was involved from 1935–1941), the show was a photo essay depicting the "human" condition on earth. Culled from Magnum and other press photo archives and other work from many of the most famous documentary photographers of the time, the display comprised of hundreds of photographs with subtitles documenting a single species; an ode to biological diversity in humanity framed by the narrative of a standard hetero-normative life cycle with the nuclear family at its center. Unlike the multi-media spectacle above it, the tempo of "The Family of Man" is slower and anchors photographs with text that mobilizes melodramatic sentiment or empathetic relationality (depending on one's view of the show). The dominant reading of Steichen's work by such critics as Roland Barthes, Susan Sontag, and Allan Sekula was as a sentimental representation of the human condition in the name of propagating American imperial ambitions – as a pivotal part of the aesthetics of empire by which the United States presented itself as the model specimen of a new global consumer species.[11] "Family of Man" arguably still operated on a model of legibility produced through the aesthetic conventions of documentary realism and the temporal narrative of a linear life pattern. As an exhibition, and in contrast to the Eameses' exhibit, it maintained fidelity to textuality as necessary for, and abetting, visuality.[12]

Algorithmic cinema

But this story of an aesthetic of imperial ambition operating through still-modern forms of sentiment, identification, and ideology may be too simple. The installation above Steichen's show had a different logic. The Eameses were not interested in life as linear progress through stages. While children and mothers appeared, and faith, certainly, "Glimpses" overwhelmingly featured infrastructure, roads, electric bridges, material pleasures such as food, nightclubs, and winking women, flying kites, and other moments of pure gesture such as the cycles and rhythms of mobility and labor as people dash into cars for work. If Steichen has been understood to present a single idea for how human beings share a world and a definitive historical and life cycle/biological time, the Eameses had a different idea – one of choice, patterning, reverberations, and redundancies.

Anticipating our contemporary arguments that their historical moment heralded both new forms of perception and novel forms of economy, Charles Eames insisted on the historical specificity of this form of visuality and media. He was explicit that this was an architecture of multi-media and *not* cinema.

The installation was, indeed, a case study in communication theory – a critical experiment in information management. The seven-channel installation was carefully timed. The flow charts made by the Eameses, miming those in computer science (with which they were familiar with through IBM, their major client), patterned the presentation.

Cybernetic vision

A central organizing principle of this display was a concept of vision linked to the communication sciences. The Eameses viewed communication theories as central to their design principles, and regularly worked with cyberneticians, such as Norbert Wiener, and corresponded with cognitive and neuroscientists of the time, such as Jerome Bruner of Harvard.[13]

These sciences at the time strenuously affirmed the idea that vision was not only autonomous but also capable of analytic, even cognitive capacity. In the same year, 1959, the classic study "What the Frog's Eye Tells the Frog's Brain" suggested that the eye was capable of cognition independent of the brain.[14] It is, until today, one of the most cited pieces in psychology and computer science literature. Whether the specific study was known to the designers of the installation is secondary to the point that it gestures to the emergence of a dominant model of the visual apparatus in the communication, cognitive, and computer sciences of the time. What is certain is that

John Whitney Sr. (a film editor and pioneer in computer graphics in cinema; see Figure 11.1) and the Eameses retroactively engaged with many of the scientists invested in this model of vision that underpinned the editing, technology, and representational logic of the multi-screen image.

The authors of the frog's eye study were all based at Massachusetts Institute of Technology in the Research Laboratory of Electronics – psychiatrist and cybernetician Warren McCulloch, neuro-scientist Jerome Lettvin, biologist

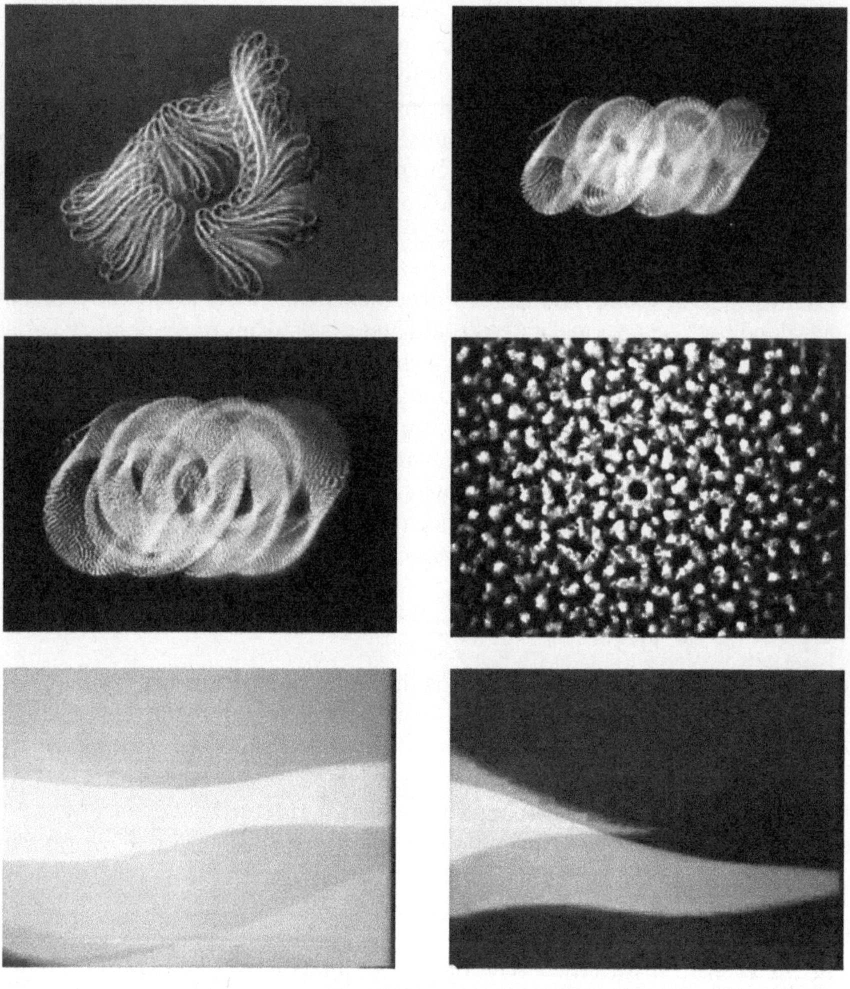

Figure 11.1 Stills from John Whitney's 1961 *Catalog*. He used his adapted servomechanisms-camera machines to produce the patterns and animations without pre-drawing the stills and to create patterned mutations and movements. This is one of the first examples of computer animation, produced shortly after "Glimpses."

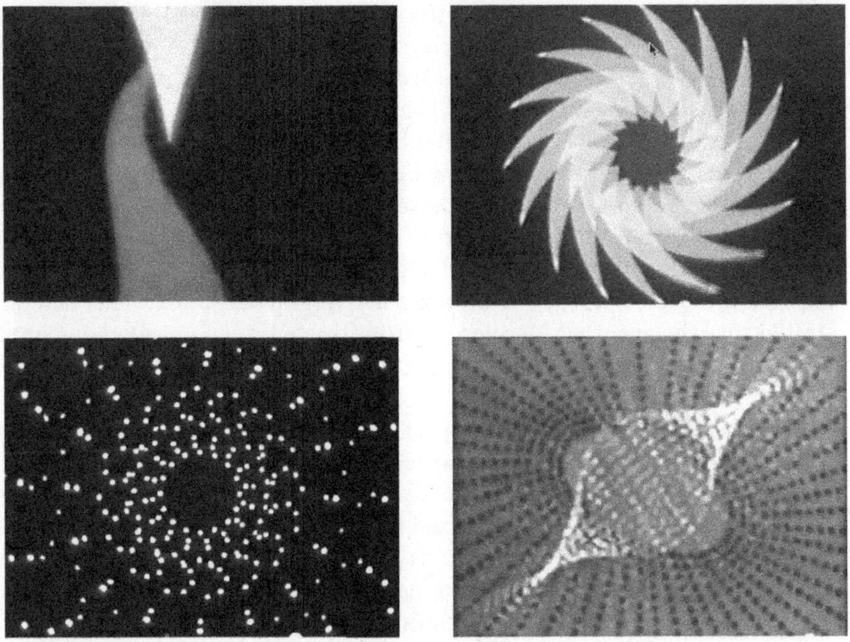

Figure 11.1 (continued).

Humberto Maturana, and logician Walter Pitts. The study opened with the following observation: "A frog hunts on land by vision. He escapes enemies mainly by seeing them. . . . The frog does not seem to see or, at any rate, is not concerned with the detail of stationary parts of the world around him. He will starve to death surrounded by food if it is not moving." These words describing the isolated optic nerve of a frog should immediately call our attention to two features: the mobility of vision and the capacity of vision to act (or hunt). Asserting a break with behavioralism and psycho-physics, the team refused to treat vision as a series of discrete separate mechanisms (a causal chain of stimulus–response behavior), imagining instead a sense built as a communication channel emerging from the relationship between that being seen and the seer. Vision itself, therefore, became a material artifact – an algorithm – capable of actions and decisions such as identifying "prey" or an "enemy."

Working within a space where signal processing and, previously, anti-aircraft defense had been prominent concerns, that vision and the identification of enemies should have been important is of little surprise. It is of even less surprise that information management would be of central interest in isolating signals. The study's framing question, therefore, was: assuming a world of information overload, how can we presume that all processing occurs in the brain? Their answer was revolutionary from the vantage point of history. Cognition, they argued, *does not* happen in a centralized location (the brain). They would prove that the management of data emerged through the networked organization of the sensation–perception–cognition system.

Their initial logic was critical. They hypothesized that the optic nerve does not transmit every piece of data (light) it contacts. This assumption reconfigured their experimental practice. Rather than test discrete stimuli, the researchers exposed an optic nerve to *variations* in light. They discovered that when the eye was exposed to stimuli simulating a moving insect or an enemy (stimuli that moved or changed from light to dark), the electrical impulse given off changed *before* ever arriving at the brain. This demonstrated that the eye itself – separated from the brain, an autonomous optic nerve on a dish – was capable of distinguishing between such binaries as prey or enemy and non-prey or non-enemy.

They concluded that "the eye speaks to the brain in a language already highly organized and interpreted, instead of transmitting some more or less accurate copy of the distribution of light on the receptors."[15] Their fellow colleague Michael Arbib summarized this finding as proof that the frog's eye could deal with universals such as "prey" and "enemy."[16] In summary, eyes were found to be Turing machines. Perception, therefore, became the same as cognition; autonomous entities – such as eyes – began the process of abstracting and processing information. This analysis opened the possibility that autonomous perception could be technologically replicated, a possibility that continues to underpin much computer science work on vision.

The emerging post-war neurosciences did not understand the image as a representation being transmitted and then translated upon arrival in the brain. Instead, they redefined vision to encompass the entire relationship structuring the act of observation: an autonomous communication channel. Such subjective perception had been found in nineteenth-century physiology and psychology, but had been a problem for scientific objectivity and knowledge; since these findings, the technological potential in nervous networks could be positively embraced. Nerves themselves, extracted from any particular body, can process and analyze data.[17]

This was an eye extended into the body and out into the world, a vision that was material and could now act on its own – flies eaten and air planes blown up, for example – a networked cognition beyond the brain and a new way to understand the differences between subjects and objects, enemies and prey. This form of vision, however, was predicated on a concept of informational abundance from which patterns are carved out and materialized.

In "Glimpses," this discourse of an algorithmic and potentially inhuman vision was directly deployed through the editing tactics and construction of the seven screen channels. The chief editor of the presentation was the avant-garde movie-maker John Whitney Sr.,[18] famous for using cybernetic concepts to make films.[19] In discussing his editing tactics he wrote of producing a "liquid architecture" that would create "structured motion that begets emotion."[20] Whitney sought a machine cinema, no longer in the realm of image and index, that would produce a new world of entropic potentials, directly tap the nervous system, and produce a mobile space. "As early as 1957," he recalled, "I had begun to construct mechanical drawing machines . . . I was not motivated to create representational images with these machines but, instead, wanted to create abstract pattern in motion . . . to evoke the most explicit emotions directly by its simple patterned configurations of tones in time. . . ."[21] To accomplish this, Whitney used the remains of anti-aircraft servomechanisms from the navy to construct machines that could produce graphics on film without filming any original drawing or live action.[22] Whitney sought an algorithmic vision that made machines authors in the production of human experience – a form of machine vision whose work did not operate at the level of the visible image but through the attenuation of the nervous system, by way of computational logic.

If the organizers of the USIA pavilion still thought in terms of images and architectures of geographical space, Whitney discussed his work in terms of harmonies, fluids, sine waves, raw patterns, and "material abstractions", implying the production of a visual-acoustic sensory environment that would transform cinema and, in fact, territory. For Whitney, vision took materiality as a process to be designed, replicated, and computationally programmed.

Machine spectators

These ideas of an algorithmic form of sight, no longer anthropomorphic in its actions, was directly refracted in the Moscow installation, with the movie star's close-up as the locus of transfer from the human to inhuman vision. The close-up, as feminist film theorist Mary Ann Doane has written, is the moment where that which is human (the face), the very sign of subjectivity, condenses into raw gesture and form. Critics obsess about the close-up as demonstrating how the cinema exceeds language while asserting the human body (the star) at its center. As Doane demonstrates, the discourse on the close-up in film theory is fraught by an effort to maintain the human at the center while attempting to repress the time and scale that make the close-up operate not through identification with the star but through the dissolution of the spectator's ego and their consumption into the image. It is the operation in cinema that most clearly demarcates cinema's inhuman and mechanical qualities while inspiring subjective desire for and identification with the screen-star-body. This joint feature (the intimacy of the face and the size of the screen), Doane argues drives a frenzied proliferation of screens, both very small and very large, in an attempt to both enjoy the cinematic spectacle (IMAX) and maintain control over the image through personalization (hand-helds). At the zenith of Doane's argument may be the failure, entirely, of the cinema as a medium in the face of another media landscape; a failure that the multi-screen installations signifying a new form of spectatorship and medium appear to embody in proliferating screens, scales, and media.[23] For the Eameses and Whitney, scale is directly material, it no longer needs to pass through the dialectic of establishing shots and close-ups; the two manifest as rhythm, not as an effort to recuperate the human body as the measure of the screen.

Utilizing the latest imaging technologies – aerial views, microscopic closeups – the purpose of "Glimpses" was thus not to align the human eye with the machine eye. There was no *mis-en-scène* in the piece. It is not that the Eameses saw everything through a camera, and that the view of the camera and that of the human being were being aligned, but rather that the camera took an autonomous role. Mechanical vision emerged because the focus on relationships and scale provided a direct conduit to experiencing emotion, as the editors of the piece imagined, rather than a conduit to seeing discrete images. This was not training in seeing like a machine, but rather in being part of one.

The Steichen show was linear, driving users to choose a pathway through a historical life cycle centered on the drama of motherhood and child development. For Whitney and the Eameses, the spectacle had to be produced through redundancy and repetition. Steichen only occasionally used landscapes, tending to focus on mid-length shots or close-ups that clearly showed emotion and allowed individuals to be singled out as unique. For Steichen, the photograph arrested time, stopped flow, facilitated identification. The images in the show worked to produce a direct association between the eye of the camera and that of the observer, and the images were hung at face level and not above spectators. In the Eameses' screens, extreme long range and extreme close-ups were the standard, usually following one after the other. Whitney's editing style followed the logic of musical scoring, favoring repetition, cadence, and harmony. "Glimpses" repeated cycles of slowness, accelerating images, and then again slowness as a way to move viewers through a "day" that was mostly about showing repetitive patterns of infrastructure and activity. This speed was paralleled musically and in scaling images, with the closest or furthest moments as crescendos – many highways at once or Marilyn's face.

The Eameses argued that images, such as highway interchanges, would be universally "familiar" (which is questionable).[24] The installation implied a format of species unification, perhaps an alternative to Steichen, grounded not at the level of the organism, but at a cellular or even molecular level, based on shared neural patterns and forms of attention rather than identification with similar subjects.

Here, a global nervous network was propagated through a conception of vision as a channel or threshold, offering people a sense they had seen something they were familiar with and that was real in a documentary sense, without necessarily allowing them to focus or fulfill a total identification with any one image. "We wanted," said Charles Eames in a later interview, "to have a credible number of images, but not so many that they [the spectators] couldn't be scanned in the time allotted. At the same time, the number of images had to be large enough so that people wouldn't be exactly sure how many they have seen."[25] This perceptual architecture insisted on an eye capable of finding patterns in vast data flows. This eye, however, could never be fully "sure": never stable, always available, as in the new epistemology of cognitive science and cybernetics, to anticipate and assimilate more data. Credibility, here, was not about knowledge but about capacities. Vision was reconfigured as a channel to be abstracted, technically reproduced, materially manipulated.[26]

Most times, Eames argued, people confuse multi-media with multi-images. In that case, this logic implied, any film is multi-media. But, he continued, the work of the Eameses' office was different: "It had not only multiple images, including the relationship between still and motion pictures, but also sensory things . . . We used a lot of sound, sometimes carried to a very high volume so you could feel the vibrations . . . We did it because we wanted to heighten awareness. . . ." Awareness, itself, is a new site of technical articulation, taking on a materiality to be the modeled and encoded as a form of media.[27] Unlike their predecessor Herbert Bayer (who had provided inspiration for the Steichen layout), the architectural diagrams from the Eameses show no observer; they render the installation as itself an eye, perhaps a cognating one. The diagrams are drawn from the perspective of the projectors. The spectator had disappeared.

Of greatest interest in demarcating a historical shift in media strategy and the management of attention was the Eameses' and Whitney's attitude to cutting and the image. They viewed relational shifts, the changing of a series or set of images together, as facilitating information exchange. Charles Eames argued that changing one image at a time was useless. Only changing images in groups of three or four so the eye could find a pattern was useful. Transfer would always start at the bottom of the screen and a number of screens would change together to allow the eye to pick up patterns. He labeled this a new form of relational editing.[28] Eames also emphasized that it was about spectator "choice."[29]

The cinematic tactics of the Eames installation, along with earlier USIA films already shown in Belgium at the World Fair in 1958, more closely related to Soviet Constructivist conceptions of an autonomous and machine vision than the classical Hollywood organization of spectacle where the spectator is comfortably aligned with the camera in the hope of reasserting the human body as the measure of the screen image.[30] But where Constructivism was linked to both behavioral psychology and utopian politics, "Glimpses' links to the cognitive and computational sciences and demonstrated different understandings of truth and human subjectivity. The notions of absolute truths or teleological progress were replaced by feedback circuits of interaction.

More significantly, montage still relied on making the operation of cinema visible, through juxtaposing idiosyncratic movements. The cuts between stills make the observer conscious of the medium's operations, and refer to the index, or reality, of cinema itself. Montage operates by constantly making visible the limits of the still and reminding the viewer of the cinematic capacity to animate individual stills. It operates through differences between images and through ruptures, rather than through continuity. In the Eames piece there are no cuts; the limit of the screen is never gestured to. Rather, the installation encodes repeating environments; shifts occur through flow where the action in an image bleeds onto another screen, finding resolution elsewhere in the installation's time and in space. Whitney emphasized flow, not editing. Mutation of images is the logic, as changes occur in waves of multiple screens, moving from one diagonal or another in the seven-screen set. This cinema was not about identification, the images move too fast, but pure affect, responded to perhaps at the level of the nervous system not consciousness. Communication, the structure and organization of the channel, not representation was the paradigm.

Whether pattern recognition and subjectivization should be considered equivalent is unclear. This was not the architecture of distraction put forth by Walter Benjamin, but it is also not an architecture where the terms of shock or identification can still be used. The individuated subject produced through normative images of the self, or propaganda assaulting psychology, does not describe this new media practice. Domesticity, identity, and psychology are all concepts applied to this installation that deserve reconsideration, or perhaps redefinition. Steichen's show, more than the Eamses', allows affiliation and identification with clear subjectivities. The forms of spectatorship facilitated by the multi-channel installation pass through the familiarity of the nuclear family and stable subjectivity to attune an observer no longer linked to norms and what Colomina labels as "enclosed" spaces.[31]

Instead, "Glimpses" offered the architecture of the network, producing a new form of spectator, simultaneously hyper-individuated and linked into a broader circuit, whose very nervous system was already conceived as a part of an interface. It is worth noting that US-administered exit polls at numerous exhibitions at the time demonstrated that the "soft" message of US cultural initiatives often originated, in fact, with the rather diffuse and unclear message of the installations. Counter to the Soviet installations that offered clear points on the virtues of Communism and the technical prowess of the state, viewers tended to articulate pleasure at American exhibitions but a lack of clarity as to the message.[32] Spectators were offered affective sensation without clear representation and without point of view.

Archive

But, if the past was forgotten, it was still stored. These installations had an archival sensibility. Absolute storage, the ability to save everything, was the unconscious desire structuring this form of visuality. However, the nature and organization of this storage system, and the identities it creates – those questions that defined the ninetheenth-century archival impulse in the work of Foucault, and so much colonial, post-colonial, and post-modern discourse – were repressed in the interest of producing these relational forms of seeing. There was no stable ontology or concern with recording here. Smooth space for global integration where the autonomy of vision, the interactivity of attention, and the absolute recordability of the world were givens.[33]

The juxtaposition of the two displays, in collaboration and competition, provokes consideration of their differing historical forms of visuality and temporality. This emergent perceptual territory relied on maintaining an ongoing tension between multiple forms of spectatorship and perception; sense vacillated *between* these two different modes of interaction. Shuttling viewers between identification and a logistics of assimilation/substitution, the affective field wavered. While the Eames display targeted circulation and arguably consumption, Steichen's show resolutely focused on the human condition separable from the material or consumer habits of its subjects. The only element unifying the two displays was the specter of the Cold War itself – the bomb.

In "The Family of Man" the atomic bomb figured as a color image only in the hardcover catalog; it had been removed from the actual installation.[34] But its logic was central to the notion of a unified but differentiated humanity. The bomb had no direct representation in the Eames installation, but the idea of a perfectly communicated affective environment spoke to averting such a possible future through the technical manipulation of the image archive. This, then, was a curious form of futurity whose imaginary was both radically nihilistic and abundantly optimistic.[35] These two operations – identification and circulation – operated consecutively to move individuals to identify with a human species now organized through communication networks. Ironically, the document filters and induces the moment of recognition necessary to embrace a form of visual experience that may no longer be about "looking" at images.

In cybernetics the tension between storage and circulation continued to animate the production of endless interfaces. Between an autonomous machine vision no longer linked to geographical space or humanity presented by the Eameses and the vision of a biologically threatened human species presented by Steichen lies the infrastructure for our current data-filled and sensory environments. This relationship between older forms of spectatorship and subjectivity, and the proliferation of interfaces, are the two poles that substantiate our contemporary aesthetic and political situation; driving a contradictory re-centering on identity and subjectivity at the same time that the human measure of the screen is effaced in the name of another discourse of direct neural and cognitive interaction and manipulation. This contradiction emerges in our present through the ubiquity of computing interfaces and global social networks, while reactionary and identity politics are resurgent in many forms.

Violence?

Twenty years later, Charles Eames had a concern: the designers of this multimedia world appeared to struggle with what their creations induced. Our own work, he recalled, "has come back to haunt us." He went on to say: "Franticness of cutting tends to degenerate the information quality. We have always been committed to information: it's not a psychedelic scene in any way."[36] They were not, Charles Eames implied, inducing non-rational mental states (although in a world where eyes can cognate, the definition of rational should not be clear). Psychedelia, after all, is affiliated with hallucinations, delusions, mysticism, forces, and powers that, while arriving from within, are understood as coming from outside the subject – which can, in a certain way, be understood as a type of violence.

At a moment when psychology and psychedelic drugs, were central mediums for control – from brainwashing to new torture techniques – we have to take seriously the Eameses' concerns about the degradation of information and the threat of the psychedelic. As one USIA officer commented, this was the greatest piece of "psychological warfare" ever waged.[37] This installation, therefore, poses critical questions about new forms of governance through media, and the infrastructures of psychology, perception, and cognition that underpin them.

These comments thus make visible the specific historical character of governance and violence. In this attentive environment that integrated our governments, our marketing mechanisms, and optic nerves, where difference—racial, national, biological – itself was deferred as a question and rendered politically impotent through consumption into an interactive architecture of hyper visualization, what still haunted this machine? What specific forms of violence do both the designers and observers of these installations express in their comments about the absence of vision and the inability to maintain a stable viewpoint, or subjectivity?

The USIA pavilion stunningly visualized the infrastructures of American life even if it made them impossible to comprehend. If anything, this pavilion cannabalized older structures of vision and gaze in the interest of consuming the possibility of evidence or of witnessing all together. It offers one possible genealogical underpinning to what anthropologist Rosalind Morris has called the "narcissistic economy" of contemporary warfare and torture.[38]

Despite the ubiquity of violent images and performances of ethical horror at torture, the present day bears no scopophilia, or pleasure in looking. Perhaps looking is not even possible. Images may no longer prompt identification or desire – whether in love, hate, or disgust. Morris argues that any concept of a social structure organizing vision and judging witnesses or participants has disappeared, becoming a "shamelessness," to use Lacan's terms, that emerges because the always recorded world is always available to be personally replayed in the very near future.

Under such conditions, circulating images do not produce evidence, proof, or emotional attachment (even if negative), but only an imperative to circulate more images. Thus, soldiers who torture prisoners continue to circulate images of their work, despite potential judgment by military tribunal, without, in Morris' words, "satisfaction" on the internet. We as a public "see" them, but only as an incentive, perhaps, to use Facebook or YouTube, and not, as one might hope, as an invitation to action or commitment to stop these actions.[39] In our ability to always feed the image of ourselves to ourselves in the near future, we make feelings of shame or remorse impossible. Morris argues that we cannot encounter difference in the field of vision; instead, the imperative to encounter is renegotiated toward an imperative for interactivity and informational circulation.

It is possible to read the consumption of racial iconography into these architectures of 1959 as servicing such a narcissistic economy of torture. In these architectures it is possible to envision that the relationship between the subject, the body, population, and territory had been severed and remixed, consuming identities and differences into a new logic of a global-species-attentive field, where histories of inequality continue to operate, but without recourse to representation or voice thus posing a terminal threat to older forms of civic life (not to mention the Civil Rights movement whose very iconography it has consumed) in the name of avoiding thermo-nuclear conflict.

I am not, however, so sure that we should be so confident in the direct relationship between this past and our future. I return to Charles Eames' concern with psychedelic states, which I read as a doubly coded concern about Communism and the counterculture; one that weakly confesses to an affiliation, while recognizing that the very architectures being designed service an apparatus for perpetual war. It is a concern for the fate of the species arising from a paradigm where dedication to "information" has replaced language.

Marx observed that "ideas . . . first have to be translated out of their mother tongue into a foreign tongue in order to circulate,"[40] and therefore the analogy between money and language, their similarities as arbitrary systems for the production of value, exists only insofar as the latter is understood as translation. Morris argues that "the bourgeois and the structuralist response to this observation has, of course, been one that fantasizes the possibility of total commensurability or translatability."[41] This is a fantasy articulated in the massive multi-channel installations in Moscow. If we are to believe Marx's older dictums on circulation and translation, we might then lose the possibility for freedom, or even a future, to the dream of perfect communication. Everything can be re-mediated or translated without change.

We must avoid this fate. We may call it an estrangement or "foreignness" from representation – but I turn to this, the core site of possibility for post-structural and other critical theories. This inability, or resistance, to perfect translation is the "source of a complex," and usually unrealized, yet possible, freedom. "It is a freedom (both the 'difference' that Marx posits as the internal contradiction of the commodity between use-value and exchange value, and the temporally defined dimension identified by Derrida's term *différance*) that is the necessary condition, or possibility for revolution."[42] It lies within this computational discovery of a vision capable of destabilizing the boundaries of the human subject. And it lies in the space between older archival orders of memory and visuality as they interact with informational regimes: the haunting that troubles and inspires Charles Eames. How we define and maintain the temporal and spatial separation between the archives of visuality and the interface is part of this struggle. Does communication and translation automatically assume homogeneity and convergence between all mediums and entities, and times? The study of the past demonstrates that the field of vision is never coherent, and always multiple. Circulation demands resistances; empires are affective and vacillating entities.

In post-war design, the tension between storage and circulation continued to animate the production of endless interfaces. But even as vision and cognition were redefined as circuits, older problems of subjectivity, identity, knowledge, authority, and signification haunted the apparatus.

In 1959, this new politics of reconfiguring population through data inundation and the reformulation of perception had already begun to emerge. A historical vantage point allows us to understand the heterogeneity of that moment, and the possibility that these new techniques could have been (but largely were not) attached to different historical and spatial configurations. At this moment, a brief interlude of détente in the Cold War, the affective field wavered between global identification, circulative consumption, and individual identity, between species being and hyper personalization. Older histories of vision and documentation supported the emerging computational and algorithmic visions. The image itself continued to multiply – computational, representational, neural. It is the work of critique in the present to explore and remember these instabilities.

Nezbutaki/forget-me-not – the flower that is not only about friendship, but also about mourning, associated not only with romance and tragedy, but also promise. Lovers in their death grew these flowers so that they may live forever in the hearts of their forlorn, thus, however, forever forestalling the ability of their partner to move on. I might also ask whether or not this is the structure of the melancholy of reactionary politics? Do we return to the longstanding ideal of the territory or of militarism as the function of the state? What will be the new politics of populations whose commensurability with subjectivity is not synchronous? The observer of such spectacles, and of contemporary networks, is, to cite Lacan, "too late," only cognizant of a self far behind the speed of contemporary networks to gather data and act on the nervous time spent at screens. This opening in time, perhaps what we label "emergence," between subjectivity and population, is both the site of technical automation and of differentiation.

This is the nature of politics, perhaps: a neuro-politics no longer attached to subjectivity, now negotiated at the level of attention and nervous networks, structured into our architectures of perception and affect, feedback providing the opening to chance and the danger of repetition without difference. Forget-me-not – the last symbol and first signal of the start of some very serious games.

Notes

1 Beatriz Colomina, "Enclosed by Images: The Eameses' Multimedia Architecture," *Grey Room* 2 (Winter 2001): 5–29
2 Jacques Rancière, *The Politics of Aesthetics: The Distribution of the Sensible*, trans. Gabriell Rockhill (New York: Continuum, 2006), 10,13.
3 George Nelson, *Problems of Design* (New York: Whitney Library of Design, 1979), 63.
4 Eric J. Sandeen, *Picturing an Exhibition: The Family of Man and 1950s America* (Albuquerque: University of New Mexico Press, 1995), 125–126; Colomina, "Enclosed by Images."
5 Charles Eames, "Language of Vision: The Nuts and Bolts," *Bulletin: The American Academy of Arts and Sciences* 28 (October 1974): 13–14.
6 Numbers are unclear, but prediction taken from "Crowds of Russians Unabated at Exhibit," *Luce Press Clipping Bureau* (July 23, 1959).
7 Pat Kirkham, *Charles and Ray Eames: Designers of the Twentieth Century* (Cambridge, Mass.: MIT Press, 1991), 343; Greg Castillo, *Cold War on the Home Front: the Soft Power of Midcentury Design* (Minn. University of Minnesota, 2010), vii–xv.
8 Cited in Marilyn S. Kushner, "Exhibiting Art at the American National Exhibition in Moscow, 1959 Domestic Politics and Cultural Diplomacy," *Journal of Cold War Studies* 4, no. 1 (2002): 6–26.
9 See also Sarah Nilsen, *Projecting America, 1958: Film and Cultural Diplomacy at the Brussel's World Fair* (Jefferson, NC: McFarland and Company, Inc. Publishers, 2011); Jane de Hart Mathews, "Art and Politics in Cold War America," *American Historical Review* 81, no. 3 (1976): 762–787.
10 I saw "Glimpses" re-enacted at the Eames office in August 2009.
11 Andrew Leak, *Barthes, Mythologies* (London: Grant and Cutler, 1994): 101–102; Allan Sekula, "The Traffic in Photographs," *Art Journal* 41, no. 1 (1981); Susan Sontag, *On Photography* (New York: Farrar, Straus and Giroux, 1977): 32.
12 Other critics have argued otherwise. Commenting that the show "looked back" upon its viewers in feedback loops, the critic Blake Stimson argues that "Family of Man" undid the ego, and forced an effort to affiliate with the Other with progressive potential. Blake Stimson, *The Pivot of the World: Photography and its Nation* (Cambridge, Mass.: MIT Press, 2006), 59–105.
13 The Works of Charles and Ray Eames, Manuscript Collection, Library of Congress, Box 11, Folder 8: Jerome Bruner 1969–1976.
14 J.Y. Lettvin, H.R. Maturana, and W.S. McCulloch, "What the Frog's Eye Tells the Frog's Brain," in *Embodiments of Mind*, ed. Warren McCulloch (Cambridge, Mass.: MIT Press, 1959), 230–55.

15 Lettivin et al., "What the Frog's Eye Tells the Frog's Brain," 230–255.
16 Michael Arbib, *Brains, Machines, and Mathematics* (New York: McGraw Hill, 1964), 32–33.
17 Jonathan Crary, *Techniques of the Observer: On Vision and Modernity in the Nineteenth Century* (Cambridge, Mass.: MIT Press, 1990); Peter Galison and Lorraine Daston, *Objectivity* (New York: Zone Books, 2007).
18 Gene Youngblood, *Expanded Cinema* (New York: Dutton, 1970), 205–207.
19 Youngblood, *Expanded Cinema*, 210–213.
20 John Whitney Sr., *Digital Harmony: On the Compementarity of Music and Visual Art* (Peterborough, N.H.: Byte Books, 1980), 43.
21 John Whitney Sr., "Animation Mechanisms," *American Cinematographer* (January 1971): 26. Correspondences with John Whitney Sr., The Work of Charles and Ray Eames Collection, Library of Congress, Folder 1, Box 117.
22 The first major cinematic release use of his machines was, perhaps not incidentally, Saul Bass' introduction to Alfred Hitchcock's *Vertigo* in 1957. Whitney, *Digital Harmony*, 83–97, 129–145.
23 Mary Ann Doane, "The Close-Up: Scale and Detail in the Cinema," *Differences* 14, no. 3 (2003); Lisa Gitelman and Jonathan Auerbach, "Microfilm, Containment, and the Cold War," *American Literary History* 19 (2007): 745–768.
24 John Neuhart, *Eames Design: The Work of the Office of Charles and Ray Eames* (New York: H.N. Abrams, 1989), 241.
25 Owen Gingerich, "A Conversation with Charles Eames," *American Scholar* 46, no. 3 (1977): 326–337.
26 Eames Norton Lectures at Harvard (1970), Library of Congress Eames Manuscript Collection, Box 217, Folder 10.
27 Gingerich, "A Conversation with Charles Eames," 331.
28 Charles Eames, "Language of Vision: The Nuts and Bolts," Eames Manuscript Collection, Script Outlines, Box 202, Folder 4.
29 Charles Eames is famous for saying: "Beyond the age of information is the age of choices," www.brainpickings.org/2012/06/15/charles-eames-quotes/, downloaded January 7, 2013.
30 Nilsen, *Projecting America*, 160–174. It should also be noted that after years of Stalin, and Socialist Realism, it's not clear how familiar a Soviet public would be with Constructivist work.
31 Walter Benjamin, *Illuminations*, trans. Harry Zohn (New York: Shocken Books, 1968); Colomina, "Enclosed by Images."
32 Nilsen, *Projecting America*.
33 Allan Sekula, "The Body and the Archive," in *The Contest of Meaning: Critical Histories of Photography*, ed. R Bolton (Cambridge, Mass.: MIT Press, 1992), 343–389; Ann Laura Stoler, *Along the Archival Grain: Epistemic Anxieties and Colonial Common Sense* (Princeton, N.J.: Princeton University Press, 2009); Michel Foucault, *The Birth of Biopolitics: Lectures at the College De France, 1978–1979* (New York: Macmillan Palgrave, 2008).
34 Fred Turner, "*The Family of Man* and the Politics of Attention in Cold War America," *Public Culture* 24, no. 1 (2012): 55–84.
35 These observations are based on the documentation of the installation available in The Works of Charles and Ray Eames, Library of Congress, Photography and Print Division, Lot 13234-1-3, Lot 13393.

36 Gingerich, "A Conversation with Charles Eames," 334.
37 Leo Bogart and Agnes Bogart, *Cool Words, Cold War: A New Look at USIA's Premises for Propaganda*, American University Press journalism history series (Lanham, Md.: University Press of America, 1995), 55.
38 Rosalind Morris, "The War Drive: Image Files Corrupted," *Social Text* 25, no. 2 (2007): 103–42.
39 Morris, "The War Drive," 103–142.
40 Karl Marx, *Grundrisse*, trans. Martin Nicklaus (Harmondsworth, UK: Penguin, 1973).
41 Rosalind Morris, "Book Review: James T Siegel. *Fetish, Recognition, Revolution*," *Indonesia*, no. 67 (1999): 163–176.
42 Morris, "Book Review," 165.

Chapter 12

Sensory topographies of wind and power in Kansas

Lisa Cartwright (text) and
Steven Rubin (photography)

This chapter is about photographing wind and the industrial wind field as a means of understanding the shifting significance and agency of these elements in a national landscape that historically has been dominated by wheat (Figure 12.1), the faded iconicity of which is apparent from its image on the flip side of the twentieth-century penny (Figure 12.2), a relic that, like wheat itself, has diminished in its stature as a material anchor of value.[1]

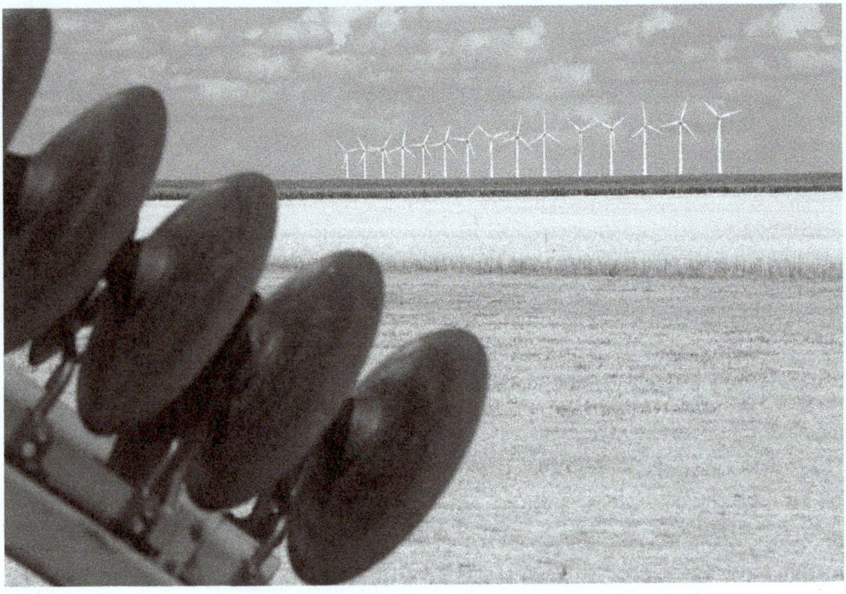

Figure 12.1 Double cropping, Gray County Wind Farm, Montezuma, Kansas.

Source: © Steven Rubin.

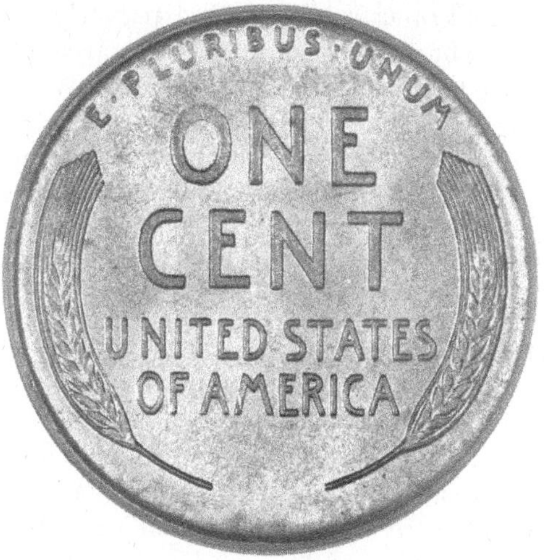

Figure 12.2 United States one cent coin from 1935.

Source: 636Buster, Creative Commons license, 2014.

Our photographs and accounts are drawn from an ongoing sensory ethnography sited in communities throughout Kansas, the Wind State, whose name was appropriated, along with its land, from the Siouan Kaw Nation, a tribe that ties the identity of its people to the wind.[2] In their introduction to the anthology *Feeling Photography*, Elspeth Brown and Thy Phu emphasize that photography, though a visual medium, is not limited to the visual register or to knowledge practices.[3] Sentiment and the experience of touch, associations of smell, and the feel of being moved by a sense of place – these are generated by condensations of place that make the landscape photograph an indexical register of history and memory, and also a topographic plan that can be appropriated for the negotiation of desire and becoming with others. Because it evades direct visibility, wind is hard to photograph – unless one turns one's back to its source, to study that which it moves. As a signifier of regional identity, wind confers this same potentiality of abstract *power to move* to the people who live in its midst. The harnessing of wind as power, however, requires new infrastructure, technologies, and practices. Photography serves as a foil for identifying the abstraction that is wind power. As an indexical medium it can be used to chart activity through and around wind power, if not wind's direct forms. Identified as a potential energy source, wind's kinetic energy since the turn of the twenty-first century has been aggressively harvested, transformed into electric power, transported, commodified, and monetized. In watching this process unfold, we witness glimpses of the infrastructural conversion of elemental power into cultural and economic power, into feelings and other resources, in a region that is still struggling to regain its corner in wheat, that now unstable agrarian coin of the realm.

This phrase alludes to the silent film classic by D.W. Griffith, *A Corner in Wheat* (1909), in which a wealthy speculator tries to corner the world market in this commodity, destroying the lives of the nation's poor, who can no longer afford to buy bread. Well known to film scholars as an early example of cross-cutting, the form of this film presents a model of the kind of intersectional mobility that occurs in the affective shifts in signification and the kinetic transference of feeling in the farm field where wheat is harvested, activities that this project seeks to disclose as they occur on the ground and in the air, between workers in grain and workers in wind. These figures inhabit the same changing topography, working in the same time and place, as they labor to yield two distinctly different kinds of crops. However, rather than featuring contradiction and difference, the chapter emphasizes the interdependent chronicities of these activities and their respective technologies. It attempts to make apparent the complex productivity of the disjunctive, inter-articulated field in which disparate kinds of activities are conducted in the intimate proximal space of a topography composed in concentrations of density and in sprawling patterns that are hard to condense into pictorial space. Informal workarounds and making do are the norm not only in the work practices of those we study, but also for us in our own use of the camera as an observational tool. Our photographic strategy has been to linger with the specificity of interaction at the very local scale of practice. The camera is a tool for rethinking the global dynamics of a national economy that embraces these elements, wind and grain, as means of shoring up unstable markets in material forms of sustenance, and as a means of engaging power. This chapter also draws attention to the presence of the camera and the photographer as witnesses situated in the field where power is negotiated.

Through photographic observation and conversations in the field with workers in wind, grain, and cattle, we have been tracking details of life in the Kansas farming and ranching topography from inside this landscape during a time that the state has taken second place in the national ranking of states with wind-energy potential.[4] During this period industrial wind developers have superimposed arrays of meteorological towers, wind turbines, utility poles, and power lines over the anachronistic patchwork topography of family and multifamily farms that compose most of the state – farms that have remained in the hands of local families for many generations, a practice protected by six decades of legislation restricting corporate takeover. But these farms have been foundering under the pressure of drought and changes in the global grain futures market. The wind industry has leased working ranch and farmland to build its sites inside this failing topography, eschewing the energy industry language of mining and drilling in favor of softer agrarian terms such as cultivation and harvesting to describe how wind is handled and processed. Yet the industrial wind facility has not been legislated as a farm – that is, it has not been held to longstanding Plains state laws and statute restricting non-local corporate farm ownership.[5] In the first two decades of this century, industrial wind technology implements have begun to punctuate the bucolic lines of this state's mythically flat topography. These new machines have not replaced the old ones (the tractors, windmills, and harvesters of the twentieth century); the old agricultural landscape has not disappeared. Rather, the topography of the state currently reflects both old and new ways of working the land. Old and new products, technologies, and practices appear together in the field.

In this project, photography serves as a means of land-use interpretation.[6] Our approach to interpreting use is not to highlight contrasts between old and new technologies, or to document the demise of old ways of working the land, but to grasp the interstitial patterns of old and new labor practices and technologies. The region still serves as iconic center of the national Grain Belt; family farming and ranching have continued in the midst of industrial wind development. But it has been newly double-branded, assuming a dual role as the national Wind Corridor even as it continues to support the older agricultural system of the family farm. What has drawn us to Kansas is the co-production of wind and wheat as they have merged together as a double-crop yield, and also as a set of interdependent mixed signifiers of regional identity and national power – signifiers embedded in a topographic landscape that supports the dual chronicities of locally owned incorporated family farms and the new high-tech energy sites, ventures owned remotely and installed and operated by outsiders. This intersectional signifying activity, the double agency of wind and grain – and, most significantly, its expression in the topographic changes over time that we observe in our tracking of sensory expression in the populated wind field – reveal the changing infrastructures of national power. Our attention ranges, then, from the visible features of place to the details of change in technological formations, embodied labor, and human–machine interactions.

Perhaps most significantly, we are drawn to those aspects of the field that make it possible to photograph wind, that element which is the region's most constant and most iconic feature, but which is only sensed and felt, evading visual perception and the very epistemological technologies of measurement and prediction designed to harness its energy for power. Our project, then, has been to use photography as a means of looking elsewhere from the abstract image of power represented in the kinetic monument that is the turbine, at what the wind enacts and mobilizes around that iconic figure, in the everyday and on the ground, and to grasp the potential of the photograph as a register of the topographic index that is the landscape itself, a slow topography of monotonous views rendered to integrate the growth of a fast-paced industry, germinating the seed of hope that was the ubiquitous wind, omnipresent in heartland images of blowing wheat.

Wheat, worry, and wind

The Plow that Broke the Plains, the classic Resettlement Administration-sponsored film of 1936 about the production of the Dust Bowl through drought and farming practices, opens with a title sequence in which director Pare Lorentz describes his film as a "picturization" of the Great Plains' transformation (Figure 12.3). The film's first image is a political map of the US with the Plains states cutting a swath down the middle. This cartographic core is superimposed by footage of blowing wheat. The moving image splits the nation, sweeping aside the two halves of the map and overtaking the space of the screen, its formerly flat, static frame brought to life by an undulating wheat field – an establishing shot of regional identity that brings the map to life. It is the status of this shot of blowing wheat as signifier of identity in place that interests us. Not just an image of wheat, this is a moving image – one in which wheat generates awe and pride. This shot serves as a counterpoint to the chaotic and destructive force of wind apparent in later shots that depict the devastating topographic impact of the Dust Bowl – shots in which this living landscape is subsumed. Our point is that it is not just the image of wheat that captures national sentiment, but also wheat's agential mobilization, from its undulations in the field to its passage through the harvester, into storage, and onto the market. Like wind, wheat must be mobilized to be monetized, and part of this mobilization is the stimulation of national sentiment about place. A heavy burden has been placed upon Roland Barthes' notion of punctum to describe this kind of sentiment that is provoked through photographs that stir a sensibility of personal history and memory, though most often the concept is applied to images of bodies and faces. Here we would invite consideration instead of his concept of the grain of the voice – that which reaches beyond the limits of language to express feeling, to grasp the body in a state of music – to interpret conditions of movement as gesture, pulsion, and timbre in landscape. We might think of the grain as itself possessing the affective capacities of voice or, to use a term Barthes borrows from Julia Kristeva, to constitute a genotext, a

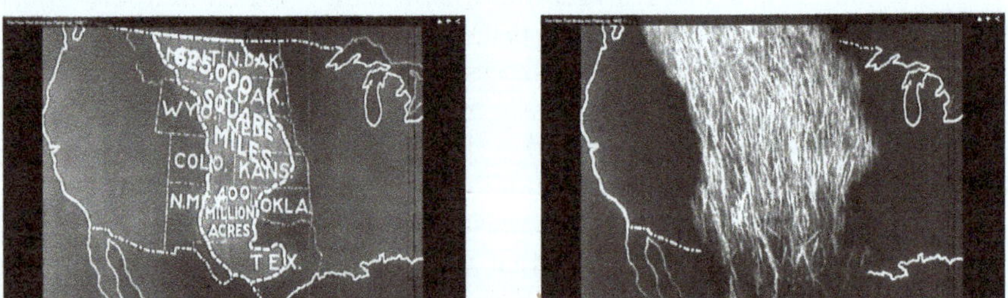

Figure 12.3 Frames from *The Plow that Broke the Plains.*

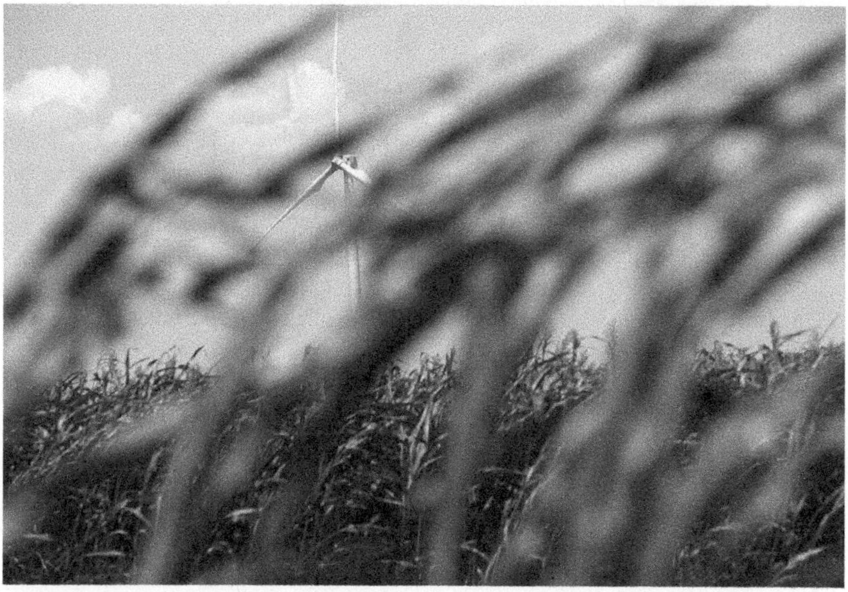

Figure 12.4 Wind turbine in a field of grain sorghum, Flat Ridge I Wind Farm, Nashville, Kansas.

Source: © Steven Rubin.

non-linguistic process that articulates ephemeral structures. In the sense of its technical and formal delivery, and its enactment of a sort of diction and breath in its "inordinate" expressivity, this undulating grain is a study in the gestural qualities more often attributed to voice or, as in Figure 12.4, a study in the oscillations of texture or timbre of the field across the static frame.[7]

But pride and awe are not the only sentiments produced in the field as affective genotext. During the years that the wind energy industry set up facilities throughout Kansas, weather and crop futures markets grew increasingly unstable, causing the Associated Press to note in 2011: "Along with the wheat these days comes a harvest of worry."[8] Worry was reflected graphically in the plain form of a state map updated monthly on the website of the Kansas Water Office. A political map void of topographic detail, this document shows county boundaries and uses color-coding to signify water conditions. In 2012, and then again in 2013, by early summer the map was filled in completely with red to signify drought emergency conditions, becoming in effect a flag of warning (Figure 12.5).[9]

The Water Office drought map signaled nothing the Grain Belt farmers did not already know. They could see their income withering on the stalk without the help of a map. It was, however, a powerful signifier of their

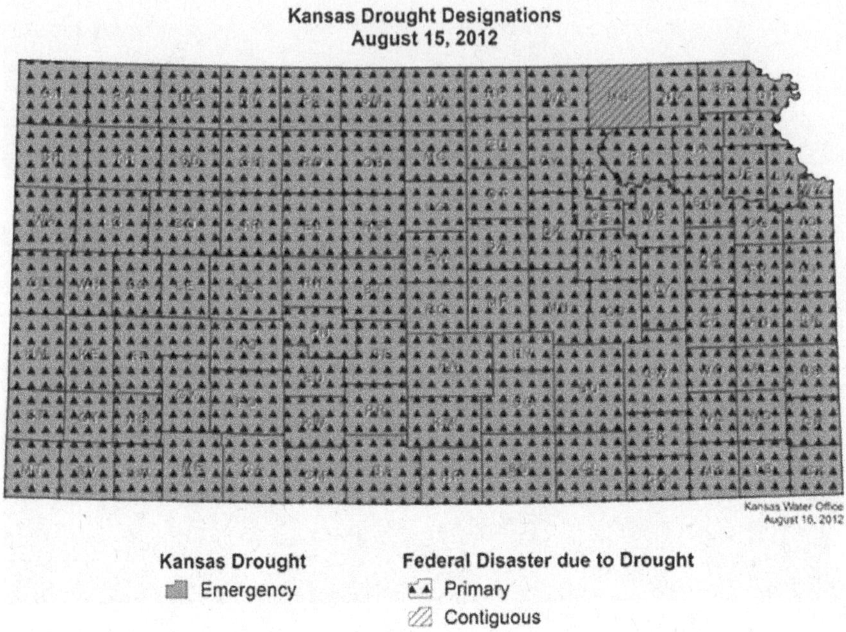

Figure 12.5 Kansas Water Office drought emergency map, August 2012.

Source: The US Drought Monitor is jointly produced by the National Drought Mitigation Center at the University of Nebraska-Lincoln (NDMC-UNL), the United States Department of Agriculture, and the National Oceanic and Atmospheric Administration. Map courtesy of NDMC-UNL.

intense worry, if not their heartfelt fear, about sustaining failure again based on knowledge from the sediment of historical memory about drought's economic impact intensified by this region's dominant pattern of land passing, along with memories, through multiple generations of a family. In the period between 2005 and 2010, droughts and floods intensified a feeling of worry that was already in the air concerning risks and potential losses in futures trading brought about by shifts toward a global market structure that produced non-convergence of timing in the storage, transportation, and trading of wheat, corn, and soybeans – grains that comprise the oldest futures contracts in the United States, and which are stored and processed using aging technologies subject to risks including grain-bin entrapment and grain-elevator explosion. These breadbasket cornerstones that historically had contributed to the securing of the region in the national economy were, in effect, upended by the transformation of the futures market as grain fatalities, drought, flooding, and tornadoes intensified the climate.

It was in the context of this climate of redoubled worry and risk that the region's farms began their transition to double-cropping with industrial

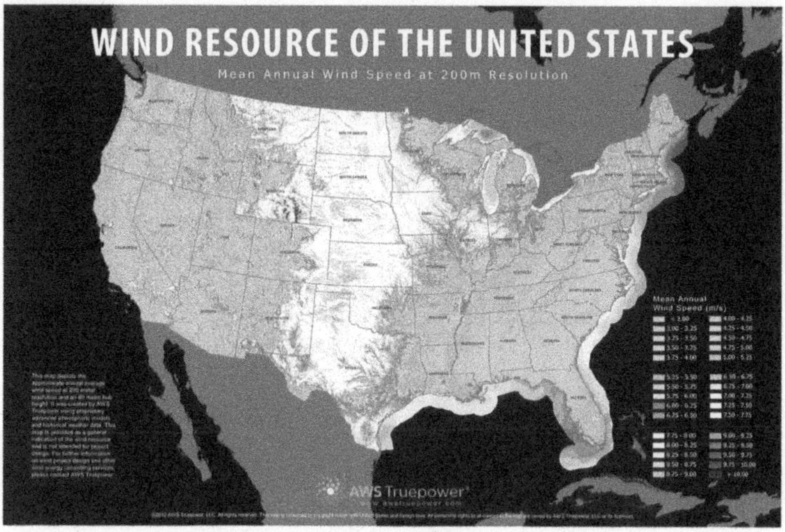

Figure 12.6 Wind resource map.

Source: Developed by NREL; data © AWS Truepower, 2013.

wind, and the Grain Belt was twinned with the new Wind Corridor. The topography of the wind region mapped almost exactly onto the old space of the Grain Belt, filling the crisis-ridden space of the Wheat State with new colors coded to reflect the distribution of specific intensities of potential wind power in each location of the state. The flat red flag of worry that was the drought map had become a topographic rainbow of hope (Figure 12.6).

The beauty ring and the office in the sky

Standing with one's back to one of the 32 wind turbines going up on the Spearville II Wind Project installation 16 miles east of Dodge City, one could look out toward cultivated farm across a boundary that divides crops from the space of the turbine. The space on which the turbine sits is a small plot of land dubbed the beauty ring (Figure 12.7). This is the designation of the circle of cleared ground to which work on and around a single wind turbine is confined. It is here from this intimate space of the individual turbine and its tongue-in-cheek beautiful ring, a tiny island of industry in the crop field, that we begin our description of the wind farm from the inside out. The beauty ring is in effect the field-lab space of each turbine. Its form is repeated throughout the field, one ring per turbine. Like a many-ringed circus, the wind site contains a multiplicity of these stage sets, cropless circles of gravel or soil each connected to an interior roadway by a gravel path that bisects

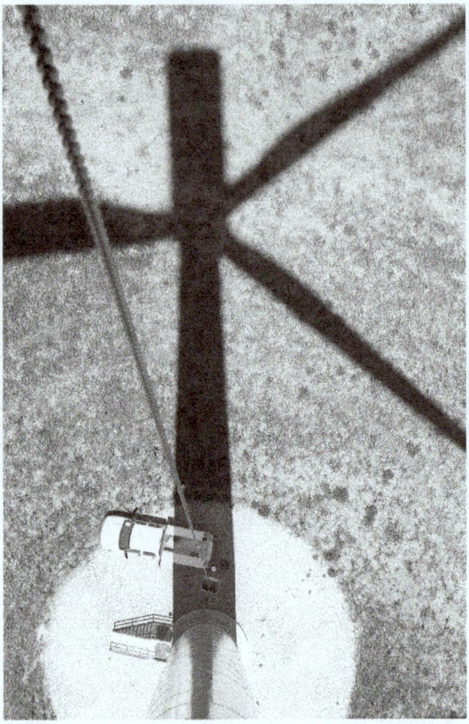

Figure 12.7 Beauty ring as seen from the nacelle, Meridian Way Wind Farm, Concordia, Kansas.

Source: © Steven Rubin.

lush crops and grazing fields, making it possible to move trucks and heavy equipment onto the ring for turbine maintenance and repair.

The term beauty rings suggests, perhaps with intentional irony, a place to be seen and admired, to be appreciated for its clean looks. Groomed and raked like a driveway apron, the beauty ring is in some places covered in aggregate or gravel with a necklace of larger stones looped around the margin, but it may also be just plain dirt, a clean-shaven sphere hemmed in all around by crop growth and from a distance above looking like the reverse image of the crop circles that grow up lush in the radius of circular irrigation systems. Chronicity of conditions such as drought's fracturing and crazing of the well-groomed space of the beauty ring gives us a slate from which to study duration and the sedimentation of concretized memories. With one's back to the turbine, a house on the perimeter of the field returns my look, its porch by happenstance angled to face this particular turbine as if the beauty ring had been designed to stage a show. Heavy parts such as the 40-ton

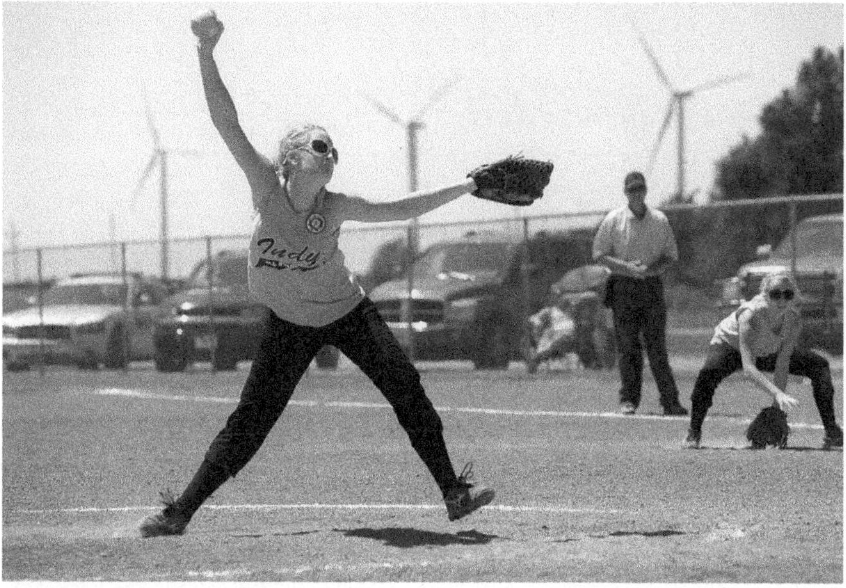

Figure 12.8 Girls' softball game near Spearville I Wind Farm, Spearville, Kansas.

Source: © Steven Rubin.

nacelle ready to be hoisted to the top of the turbine wait tethered to anchors lest their aerodynamic forms alight in the constant wind that has put the completion of the installation of 32 turbines on hold even as crews feel the pressure of finishing in time to come in under the wire of expiring government subsidies. Spearville II is the second major wind site installed on this side of town. At the time of its assembly, another company was installing a site on a network of farms across the highway that runs through the middle of the town. Designated the City of Windmills in the early twentieth century, Spearville is an agricultural community in which farmers now live and work in the shadow of turbines that have risen up like trees on the landscape, introducing their iconography not only to the farm, but also to the town's institutions of worship and learning (Figure 12.8).

But in the field lab that is the beauty ring, the wind industry has carved out a private space for itself inside the ongoing life of the community. The repetitive patterns of tractors and combines as they trace the arc of crops (Figure 12.9), and the spiraling circles of horses and pickups engaged in cattle roundups, must be adjusted to avoid crossing over onto the pristine field of the turbine's ring.

But from the crop line the frame of a door is barely visible in the smooth surface of the tower (Figure 12.10). Following service workers into this door,

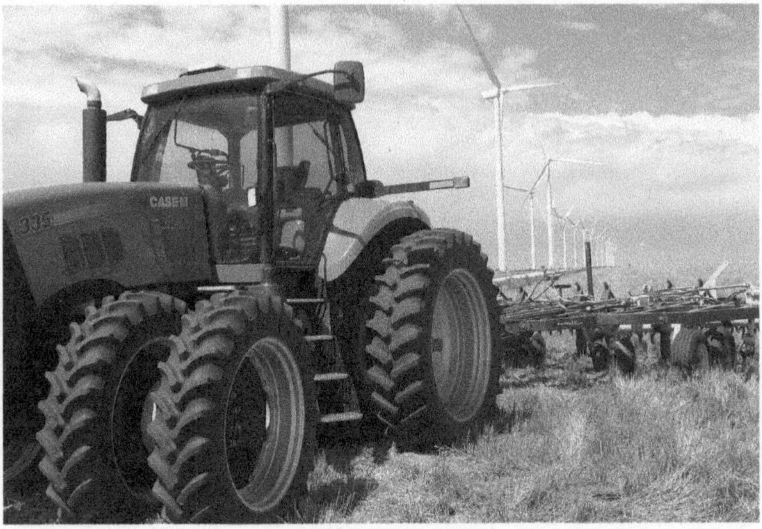

Figure 12.9 Tractor with disc harrow, Central Plains Wind Farm near Marienthal, Kansas.

Source: © Steven Rubin.

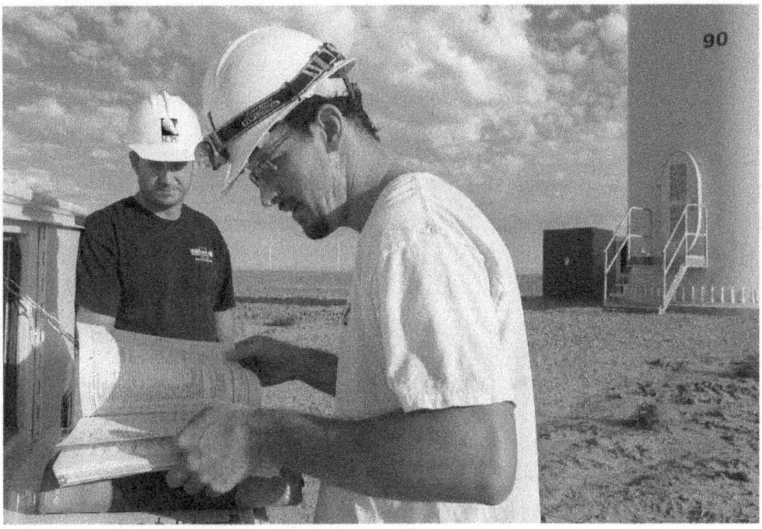

Figure 12.10 Wind technicians outside maintenance access door, Spearville II Wind Farm, Spearville, Kansas.

Source: © Steven Rubin.

one feels the intense rush of air pressure and noise from fans and motors and the array of equipment inside. Although the tower may appear to be nothing more than a support for the nacelle and the blades, its confined interior space is in fact a critical center of operations – an office with an array of machines requiring attention under conditions of noise, heat, low light, and the pressure of very close quarters (Figure 12.11). This is the projection booth from which the movements of the blades above are controlled and regulated, and the first of many locations deemed spaces of hazardous communications, as well as the first of multiple locations where wind energy will be stored.

The ladder visible inside the turbine leads to the top, where one finds the nacelle, the operating system in which generating components are found and onto which the blades are mounted (Figure 12.12). Here one finds another office, even more private and more central to the communications system than the one below. Dubbed by industry marketers the new "high-tech office in the sky," this is a new white-collar workplace replete with computer systems – implements requiring maintenance and repair at 400 feet in the air.

Field rescue: suspended agency on the ground

Across the state in Concordia, Cloud County Community College (CCCC) operates the first approved wind energy technology degree program in the state out of a strip mall, next to the Dollar Store (Figure 12.13).

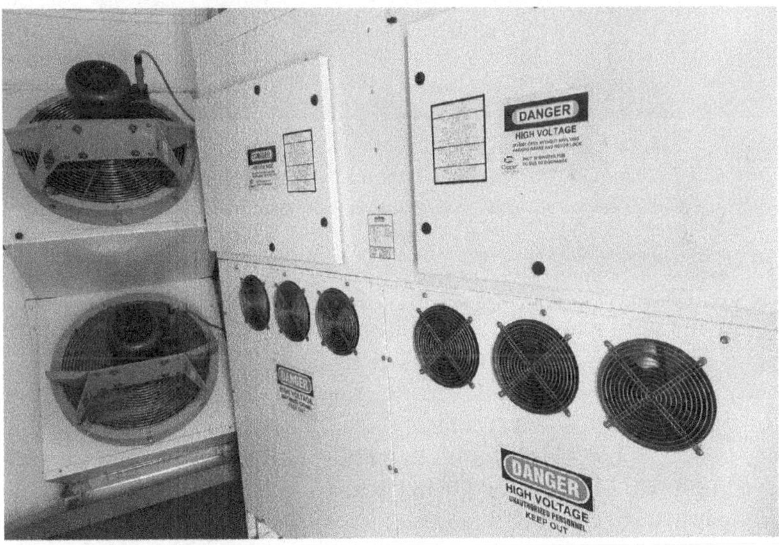

Figure 12.11 Interior of the turbine tower at base level, Flat Ridge Wind Farm, Nashville, Kansas.

Source: © Steven Rubin.

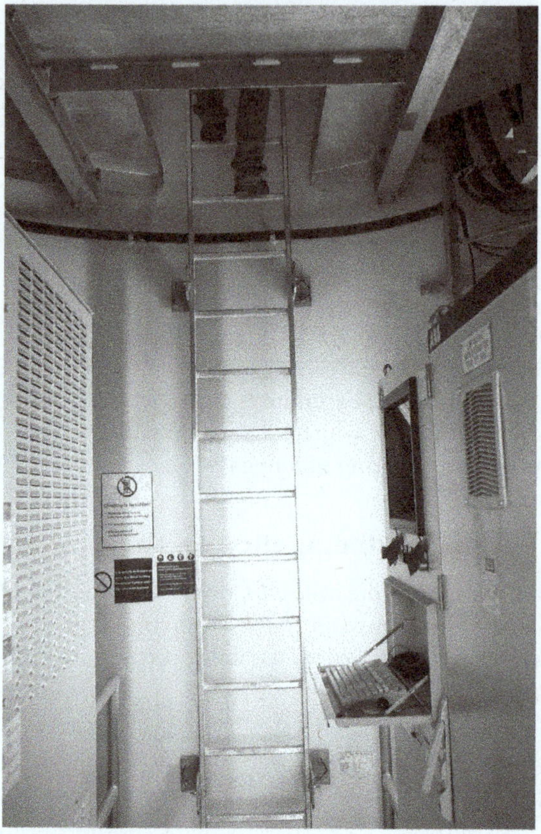

Figure 12.12 Base level with ladder and control box inside the turbine tower, Spearville II Wind Farm, Spearville, Kansas.

Source: © Steven Rubin.

The industry requires wind technicians skilled in tower climbing, working in closely confined spaces at extreme heights, rescue of others, and self-rescue. Performing these tasks at extreme heights and in confined spaces is just part of a job that also requires skill in blade repair and generator servicing, as well as knowledge about harnesses, ropes, ladders, and strategies for arresting a fall. Members of a field training course in wind service technology led by instructor Bruce Graham suit up in climbing gear to receive training in these techniques from expert skilled mountain climbers and veteran fire fighters (Figure 12.14).

To train for the job of performing servicing and rescue at the top of the tower, these trainees dangle in suspended agency from nacelles planted

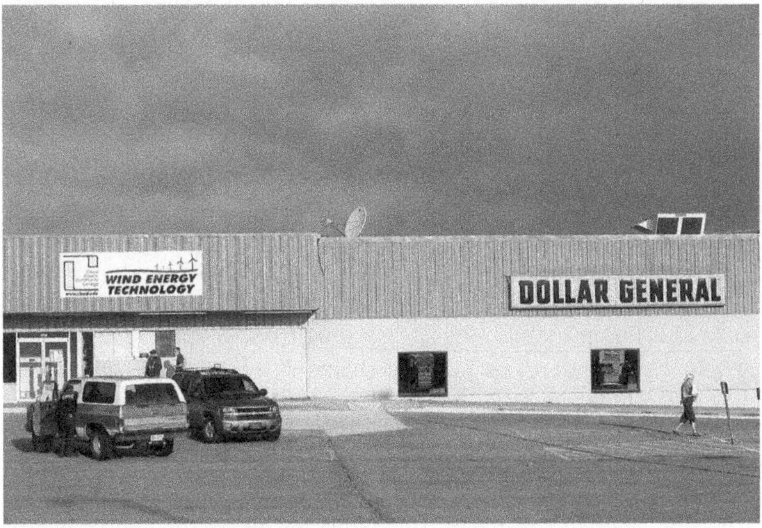

Figure 12.13 Storefront, CCCC wind energy technology program, Strip Mall, Concordia, Kansas.

Source: © Steven Rubin.

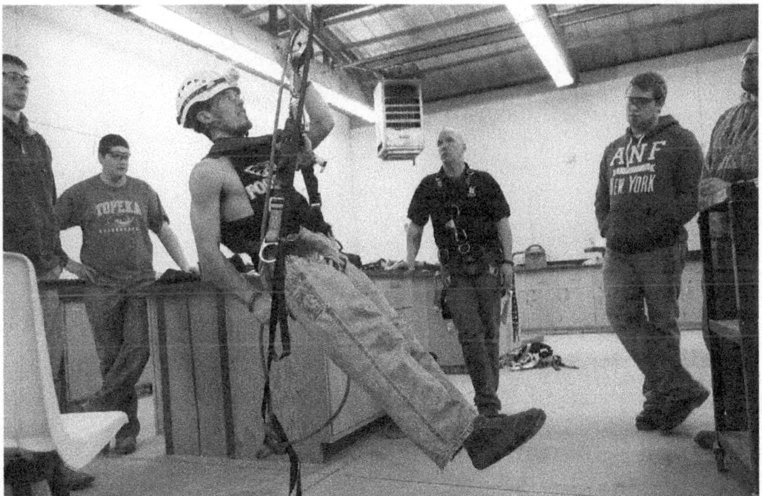

Figure 12.14 Student in safety training, CCCC wind energy technology program, Concordia, Kansas.

Source: © Steven Rubin.

directly on the ground, a space from which trainees simulate ascent and tech service fully suited for the climb. They will also practice rescue of peers in simulated states of incapacity, should they ever have to rescue someone from tech work in the "sky office" gone awry. The training session is as much as exercise in acceptance of vulnerability, dependency, and accommodation of weakness as it is training in strength, leadership, and skill (Figure 12.15).

Male bodies are entangled with technology here, but they are also entangled in gear and in mutual responsibility for and dependency upon one another. These intersubjective performances constitute a kind of industrial theater of the everyday labor of a monumental task. But this work is brought down to scale, grounded in the same sorts of concerns as work in the welding industry and the grain silo – high-risk jobs from which some of these students were laid off after 2008, just as the wind industry was taking off. Like welding and silo labor, wind technology jobs that involve aerial labor entail acceptance of the need to recognize one's limits and rely on others in times of tremendous physical risk. In Figure 12.16, members of the CCCC field class look on from behind as a student, hard-hatted and harnessed for the climb, casts his eyes pensively to the ground while his instructor, who will lead him on a rescue simulation, adjusts the rope that will secure him in the event of a fall. A look of pensive concentration and perhaps concern has passed over

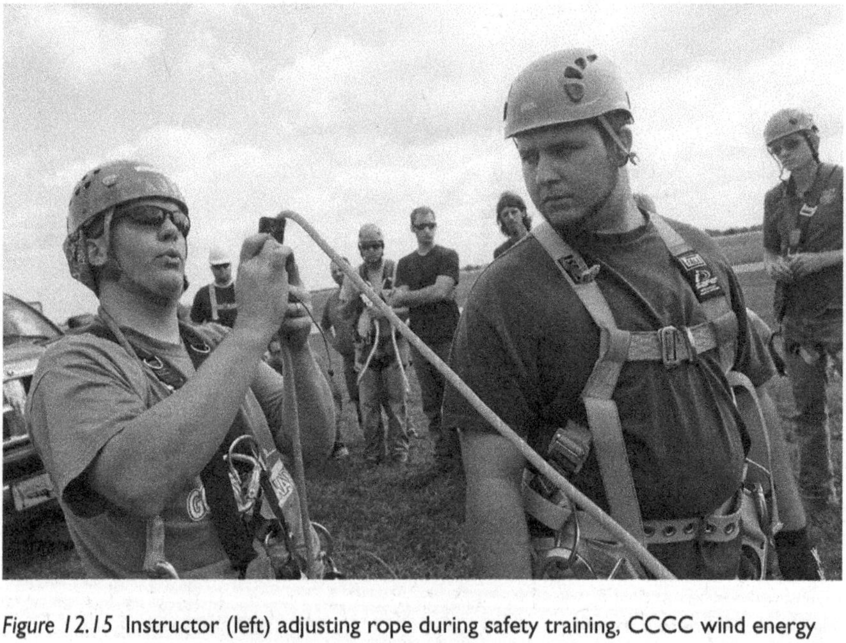

Figure 12.15 Instructor (left) adjusting rope during safety training, CCCC wind energy technology program, Concordia, Kansas.

Source: © Steven Rubin.

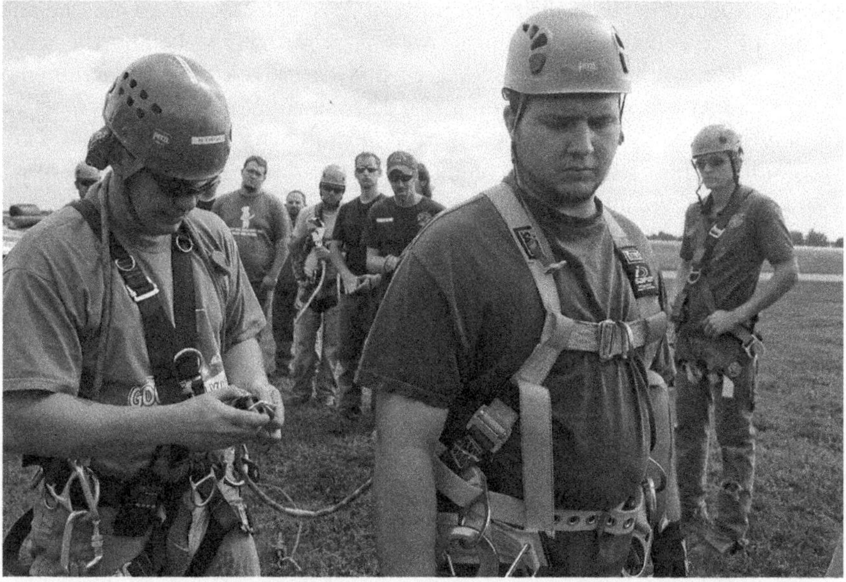

Figure 12.16 Instructor (left) with students, CCCC wind energy technology program, Concordia, Kansas.

Source: © Steven Rubin.

the novice's face as he turns his head in the direction of the partner who he cannot fully turn to see, as if he feels the pull of his instructor's hands adjusting the cord that will connect them on the climb. What we wish to emphasize here is the mutual comportment of care and dependency, and the attention to the details of the rigging that connects the two bodies. The student who casts an equivocal glance over his shoulder is caught in a state of vulnerability, even as he is about to train to become a hero – that is, to simulate scaling a monumental height in order to rescue a critically injured peer. The huddle of men standing behind him look on at the scene of preparation with expressions of equally pensive and equivocal attention, knowing that they too will have to perform in the role of the rescuer modeled for them here, and knowing that they too will feel vulnerable and slightly embarrassed even as the harness conveys some degree of later potential for bravado and prowess.

The instructor goes through the motions of adjusting the pulley and demonstrating its use, performing the smallest tasks with flourish to ensure that his audience follows the drill.

When the rescue simulation is enacted, the performance takes place a few feet off the ground, with one of the students playing the role of an incapacitated worker suspended in the air, his body limp like that of a sleeping child inside his harness. The instructor gets in close to the student who simulates the rescue, climbing over the top of the nacelle to demonstrate to him the fine details of hand coordination on the rope as someone else behind the scene handles the rope that lowers the rescuer's body to the unconscious victim who he must convey to safety (Figure 12.17). There is humor in this play image of the lifeless body, the hero, and the instructor as director, in part because there is, of course, no catharsis in a "high-elevation" rescue on the ground of an "unconscious" victim who fakes his incapacity. This is risk as slapstick physical comedy. That these are men who are for the moment ineffectual at their critical tasks lends a poignant and private edge to these images.

Our point here is to draw attention away from the rhetorically overblown image of the lofty high-tech office in the sky and the lone skilled technician who labors there, part mountain climber part geek, to the grounded intimacy of workers who engage in close proximity and in maneuvers of protection and care upon their own bodies and the bodies of others in confined and intimate spaces inside and above the field, in the beauty ring and on the body of the nacelle, well out of sight of the observer from the side of the road. The degree of interdependency among these men is rendered greater by the intervening presence of gear, which must be adjusted, touched, and tightened or removed by others, and the smooth body of the nacelle, which becomes a multi-dimensional platform against which to perform the complex acrobatics

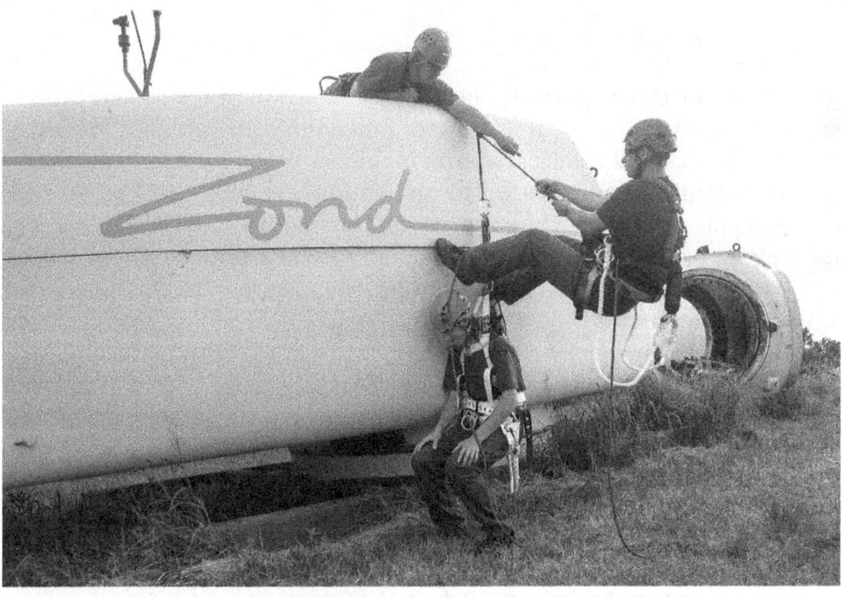

Figure 12.17 Rescue simulation, CCCC wind energy technology program, Concordia, Kansas.

Source: © Steven Rubin.

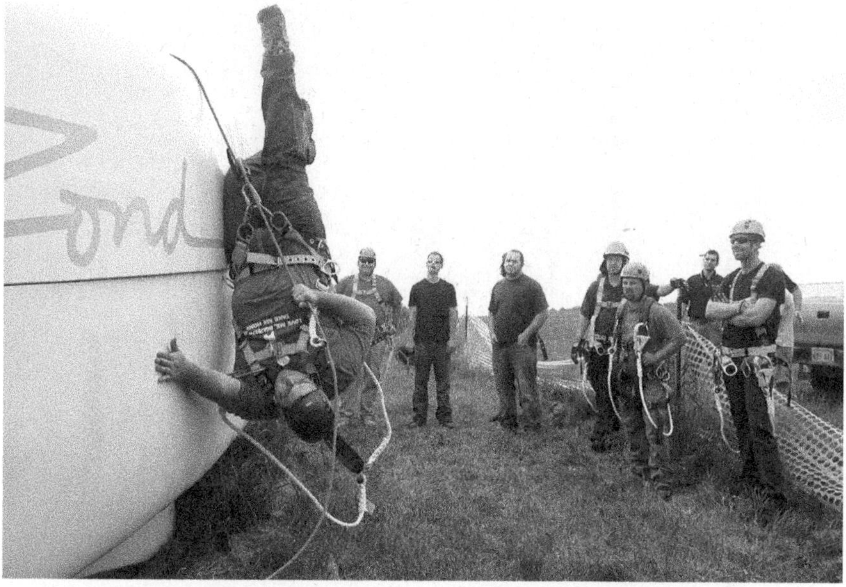

Figure 12.18 Unsuccessful repel during training, CCCC wind energy technology program, Concordia, Kansas.

Source: © Steven Rubin.

of the rescue, making the analogy to the circus ring all the more relevant. Like circus acrobatics, the conditions of spectacle are grounded by the close and intimate performance of bodily care and support, and a level of skill with technologies of bodily protection – a kind of prosthetics of the profession that facilitate the successful performance of tasks that entail high levels of risk (Figure 12.18).

Analogies to the work of the fire fighter and the mountain climber are also obvious given the professions of trainers in the wind safety field, but perhaps the most relevant comparison is to the work of walking down the grain, a practice that involves intense collaboration and awareness of the safety of one's peers inside and around grain bins, silos, and elevators (Figure 12.19).[10]

The matter of scale is critical in a field in which technology extends the topography of practice to 400 feet above the ground. Yet inside the private offices and classrooms of the wind industry, the technology is made smaller than life, handled and displayed in miniature, as icon and as teaching model (Figure 12.20). In its smallness and its cuteness, the scale model embodies the banal underside of the charged new field, its interior where the labor begins to look less like the sweeping change billed in literature about the new energy field.

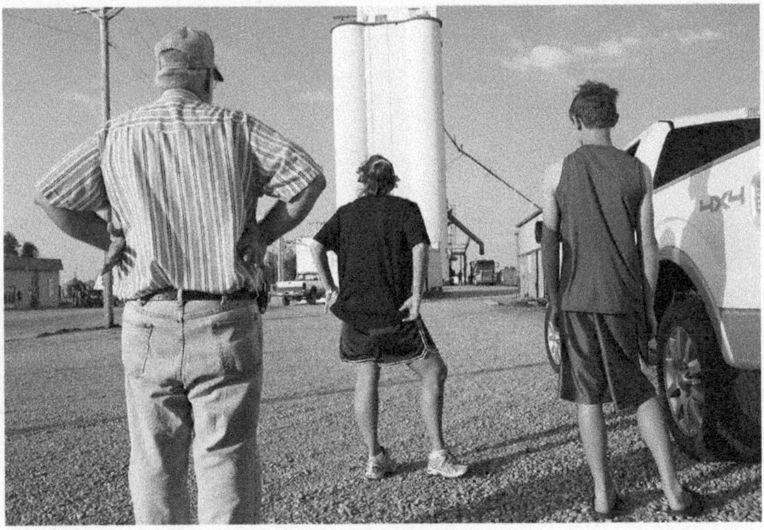

Figure 12.19 Grain elevator with onlookers, Spearville, Kansas.

Source: © Steven Rubin.

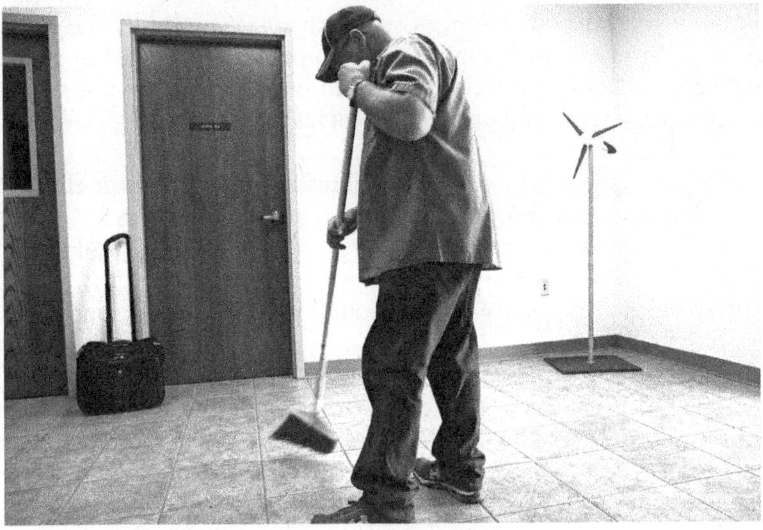

Figure 12.20 Employee, BTI Wind Energy, Greensburg, Kansas.

Source: © Steven Rubin.

Conclusion: seeing topographically

We have shifted our focus from the landscape to the topographic details of bodies that labor above that land. It becomes necessary to make a case for reading bodies topographically. Saskia Sassen, in an essay in which she interprets the contemporary city, proposes that to examine a region in terms of its built topography is increasingly an inadequate approach to understanding digital globalism. Topographic representations, she proposes, fail to capture the fact that the components of topography one sees from the standpoint of place are likely to be spatializations of larger global power projects. Capturing the pattern and scope of the larger systems through which power is distributed is difficult if not impossible to achieve in locally situated observational research about place. The wind turbine field is no exception to this logic – it is a digital landscape that is global in scope, with power grids extending out of frame, out of any possible photographic field of vision (Figure 12.21).

Yet this very idea that power has dispersed geographically through the digital, and therefore cannot be observed from the standpoint of any given place, is one of the very suppositions that Sassen has challenged throughout much of her career. She has consistently shown the value of identifying circumstances of place in the project of tracking and interpreting the distributed networks of power in digital worlds. If we follow the transmission line

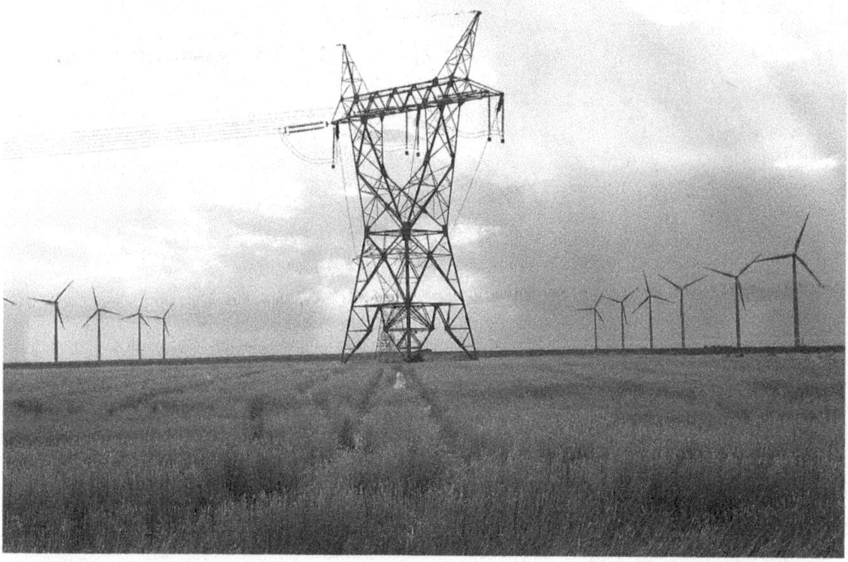

Figure 12.21 High-voltage transmission lines, Spearville I Wind Farm, Spearville, Kansas.

Source: © Steven Rubin.

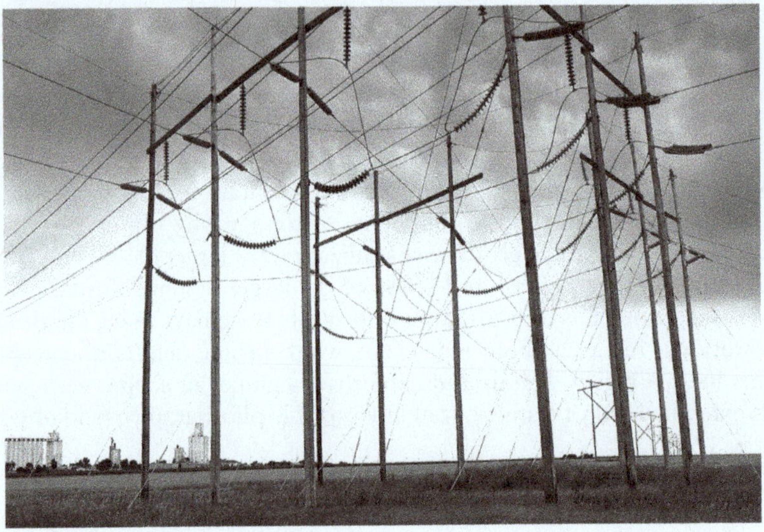

Figure 12.22 Power lines, Spearville, Kansas.

Source: © Steven Rubin.

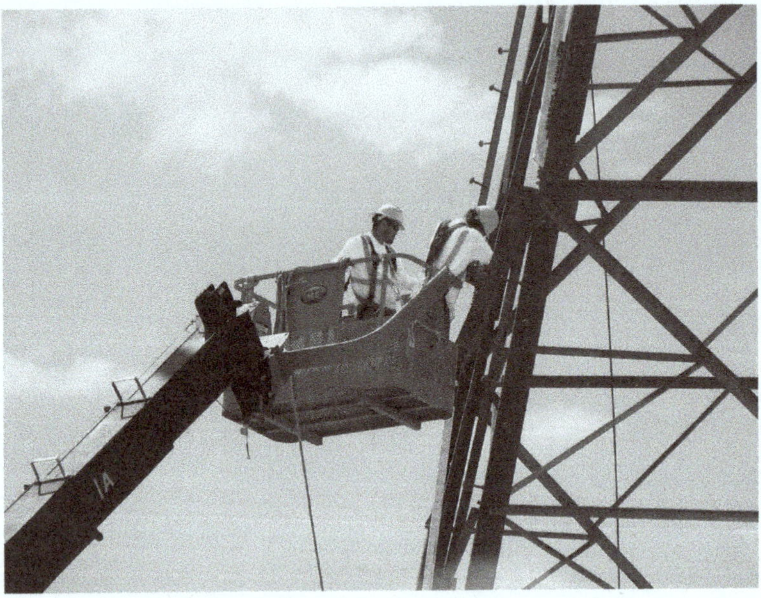

Figure 12.23 Bringing Ironwood online, Spearville, Kansas, 2012.

Source: © Lisa Cartwright.

(Figure 12.22) back to the installation of the wind site, we find the lineman suspended in action, a familiar figure whose labor prefigures the wind technician who services the wind office in the sky (Figure 12.23).

But this still does not get to the matter of bodies and topography. Here we wish to bridge the concept of the topographic with the method of observational visual ethnography. By flagging topographic representation, Sassen indirectly problematizes both direct observational methods and the use of tools of visualization – tools such as geographic information and mapping systems, and video and photographic cameras used to study the social dynamics of place. These are tools of trade for ethnographic and observational work in anthropology, sociology, and communication studies as well. We are particularly interested in Sassen's problematizing of the topographic for two reasons. First, in observational scholarship the question of "seeing place" is usually only secondarily about how the researcher him- or herself enlists seeing as an observational tool of the trade. Rather, the concern is usually how seeing and topographic construction are themselves lived and practiced by technicians and workers observed in the field. How people see in the laboratory, how a field is shaped by people who use and reconfigure it, are the urgent matters at hand that are described in laboratory studies and in science and technology studies (STS) and medical anthropology that is about visual practice (Morana Alač's *Handling Digital Brains*; Barry Saunders' *CT Suite: The Work of Diagnosis in the Age of Noninvasive Cutting*, or the work of Annamaria Carusi on biological models and simulation in data-intensive computational science). Capturing the aspects of mobility and change in a topographic system is one of the challenges faced by the STS researcher who studies the topographic elements of place, whether he or she observes a built environment and its blueprints, maps, and photographs, or a laboratory and its use of imaging practices and audiovisual texts. Attention to mobility has in fact shifted the dynamics of visual studies in science from work on the image to work on technologies and practices of imaging and mobility in the last decade – and here, in the previous photograph of line workers, we see laborers rendered mobile in a spectacularly cinematic device, a technology that mobilizes their gaze and their hand in the work of installing and making live the cables that will carry stored wind energy offsite and out of state.

The risky proximity of these workers to lines of power is managed by an infrastructure of safety that mimics the classic high-post fence of wagon days, and which serves as a barrier – should the wires fall, they will be caught by the suspended log rather than coming into contact with bodies below them (Figure 12.24).

Topographic representations, we surmise, are among some of the limited resources we have access to if we want to empirically study the precise ways in which the multifarious lines of flight in networked power systems literally touch ground and are handled by the human subjects whose labor is required to reroute power elsewhere, to other sites.

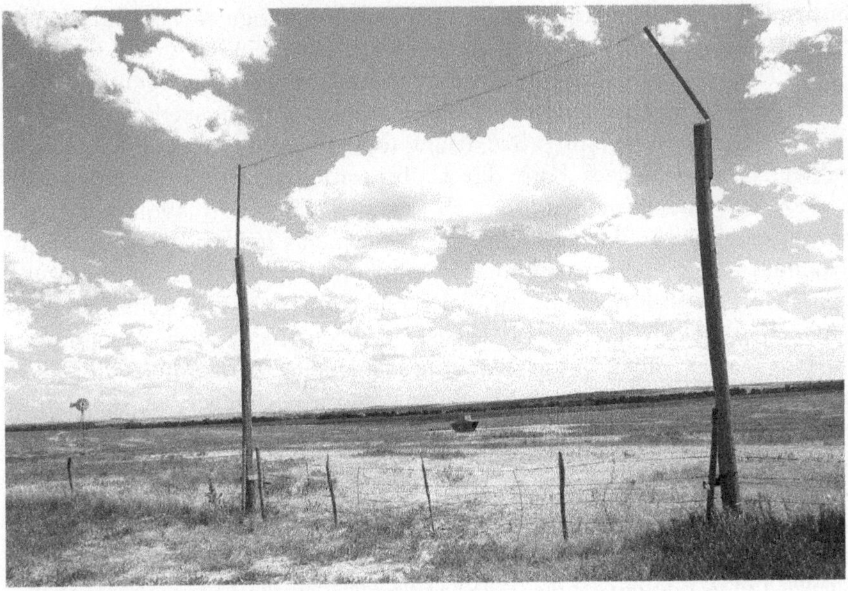

Figure 12.24 High-post fence, Wright, Kansas.

Source: © Steven Rubin.

A small fact about Kansas roads is that wagon-wheel ruts carved from early interstate trade are still visible along parts of the Santa Fe trail, embedded in untilled land in the diminishing parcels where prairie remains. We failed to document this mythology. On the drive between wind sites, though, we frequently encountered road-repair crews at work, sometimes covering over with dirt and gravel the cement undergirding designed to support trucks carrying heavy parts, restoring the reinforced highway to its rural look and feel.

We also encountered truckers caravanning with loads of heavy parts destined for wind farms under construction around the state. It's well known among the truckers who contract with the wind companies that the heavy parts that make up wind turbines can be transported across this particular state with relative ease because of the effort the state has put into roadway reinforcement to encourage wind industry development here.

On any given day one can see nacelles, blades, and towers rolling around the state like balls on a pool table (to borrow the words of a wind engineer from North Carolina bringing online the Spearville II wind site) (Figure 12.25).

Transporting these parts is no small matter. A part that weighs on average 50 tons, the nacelle is the housing that holds all of the wind turbine's generative components. It is a component in a global network that in the first half of the decade began with production overseas and continued with transport by

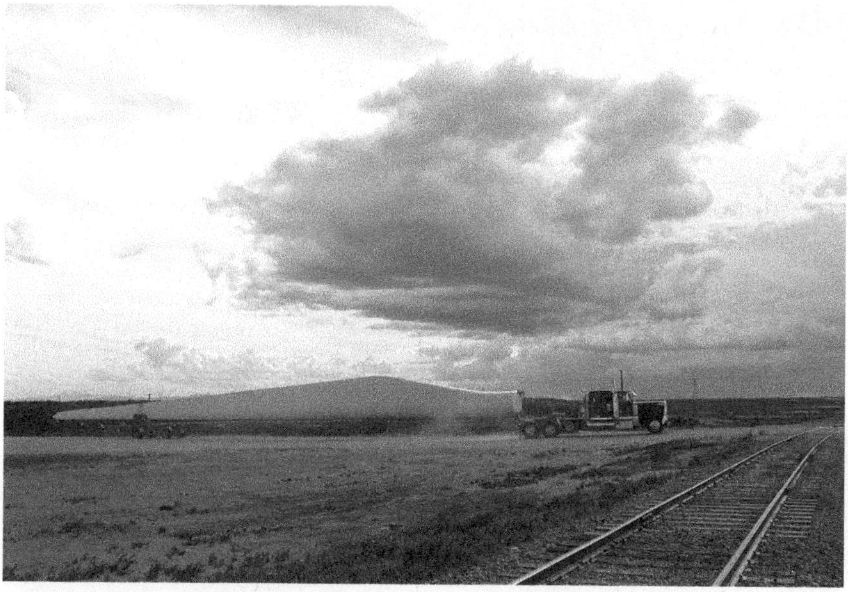

Figure 12.25 Truck with wind turbine blade, Carr, Colorado.

Source: © Steven Rubin.

ship to western ports and then by truck through secondary highways rebuilt to bear the weight of these parts.

In 2009, Steven Rubin encountered a trucker who had just lost a load in Wyoming when he attempted a U-turn on an unreinforced road. The 40-ton nacelle tilted off of the truck bed, its edge sinking into the asphalt blacktop, embedding itself like some strange meteorite striking the earth (Figure 12.26). When Steven got to this scene hours after the event, the trucker was still waiting for help to arrive, haplessly contemplating his own personal mark on the topography, an indexical mark that signified his own private economic disaster.

Of course documentation of topography and its incidental alterations has a distinct photographic history. In 1975 the exhibition "New Topographics: Photographs of a Man-Altered Landscape" featured photographs (including some of the typological series of Bernd and Hilla Becher) that were received with bemusement for their meticulously artless lack of overt style and their aversion to what were then the dominant tropes of beauty in landscape photography.[11] Considered one of the most influential American photography exhibitions of the century after Edward Steichen's 1955 "Family of Man," the show is reported to have left its early audiences bemused and disappointed. Sentiment was pointedly avoided in these works in favor of the clinical gaze Barthes famously branded as *studium*.[12] The photographs

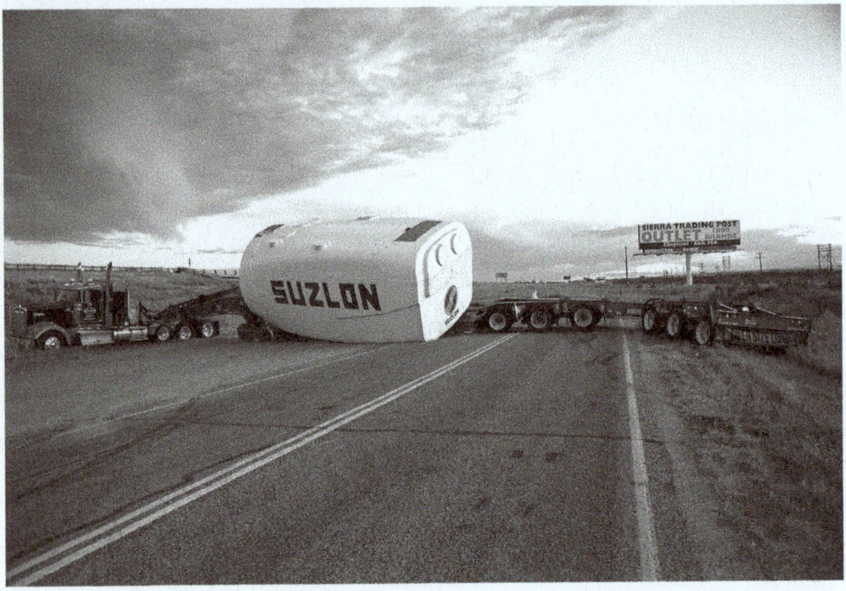

Figure 12.26 Spilled nacelle, Pine Bluffs, Wyoming.

Source: © Steven Rubin.

included in the legendary exhibition conveyed rich evidence of the everyday practices, from incidental to planned, that had configured the new American landscape, most notably the remote but highly technologically developed spaces beyond suburbia where dams and oil rigs were configured. The absence of bodies in these photographs was as disconcerting to viewers as their formal lack of sentimentality. As curator William Jenkins explained, the new topographics style was "anthropological rather than critical," its proponents running interference against the pictorial romanticism that was a legacy of the Weston era.[13] Perhaps most disconcerting to these audiences was the fact that most of the photographs had no face, depicted or implied.

We have tried to interpret wind power in light of its status as a topographic infrastructure that, like grain and also like television, links the agricultural region and its farmers of wind and wheat into a national and global power network of major consequence beyond what can be seen from the standpoint on the ground. At the same time, we have tried to convey the matter of the topographic from a standpoint on the ground that takes into view the bodies that traverse this field, expanding on the aspect of the topographic that asks us to attend to detail and contour, to bring into precise relief the matter of emplacement on the land and the conditions of land use as they are enacted by hand and on the ground, often out of sight of the casual observer who passes through.

A condition sometimes glossed over in accounts about the wind industry is that in places such as Kansas, wind farms are embedded in a working landscape that is uniquely sensory and emotional. The industrial wind farm is a corporate entity literally stitched together across a patchwork of farms on family land, land leased by farmers to corporate wind outfits but that continues to worked in conventional farming and ranching ways by multiple generations of families and by children, by individuals with an intimate attachment to the land. Looking at everyday labor and life where wind and crop workers intersect, and looking from the standpoints of the affective and sensory experiences of workers in wheat inside landscapes newly defined by the corporate double-crop of wind, thus also counts as seeing inside the contemporary apparatus of power, which extends not only to the spaces of the community, but also to the farm as the private space of the family, to the future imagined or dreaded by the child, and to the space of imagination and play that is always contiguous with the spaces of commerce and industry (Figure 12.27).

What we have documented is not the demise of the pastoral or the malevolence of the high-tech machine introduced to the agricultural garden,[14] but rather the interstitial conditions and practices that have unfolded as an energy industry has been introduced in the image and in the place of the operational industrial family farm, with turbines and power lines mixing with the silos,

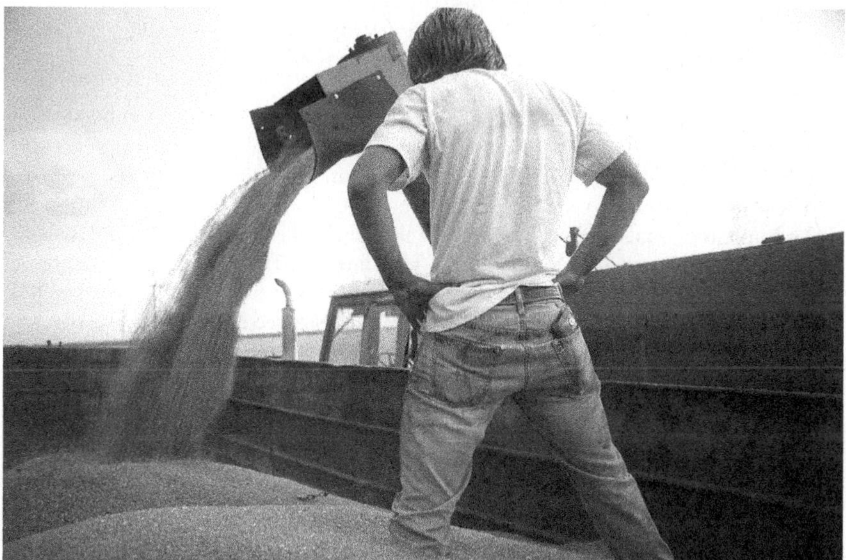

Figure 12.27 Teenage worker in grain wagon during wheat harvest, Spearville, Kansas.

Source: © Steven Rubin.

grain elevators, tractors, and combines that constitute the interlocking patterns of life inside this agricultural topography.

Notes

1 The Lincoln wheatback penny was produced between 1915 and 1958. The (standard non-collectible) penny now holds a material value less than its symbolic value (as has always been the case with paper money), because pennies now cost more to make than they are worth. In 2012 the cost of making one penny was estimated at 2.41 cents, prompting a Google+ Hangout with President Obama on the topic of retiring the coin. On coin value, see Georg Simmel, *The Philosophy of Money*, trans. Tom Bottomore and David Frisby (Boston: Routledge, 1982), 146–160; On food and money as signifiers, see Catherine Gallagher, "Raymond Williams and Cultural Studies," in *Cultural Materialism: On Raymond Williams*, ed. Christopher Prendergast (Minneapolis: University of Minnesota Press, 1995), 313–316. On the significance of valuelessness, see Brian Rotman, *Signifying Nothing: The Semiotics of Zero* (London: St Martin's, 1987).

2 See the official site of the Kaw Nation: People of the South Wind, http://kawnation.com/?page_id=72. The Siouan origin claim for the word "Kansa" is contested. It was made by William E. Connelley in *A Standard History of Kansas and Kansans*, Vol. 1, (Chicago: Lewis Publishing Company, 1918), 196. An overview of the debate that ensued from Connelley's claim can be found in William E. Unrau, *The Kansas Indians: A History of the Wind People, 1673–1873* (Norman: University of Oklahoma Press, 1986), 6–12.

3 Elspeth Brown and Thy Phu, "Feeling Photography: An Introduction," in *Feeling Photography*, ed. Elspeth Brown and Thy Phu (Durham, NC: Duke University Press, 2014), 1–28.

4 American Wind Energy Association AWEA Fourth Quarter 2012 Public Market Report (AWEA), January 2013, http://awea.files.cms-plus.com/FileDownloads/pdfs/AWEA%20Fourth%20Quarter%20Wind%20Energy%20Industry%20Market%20Report_Executive%20Summary.pdf.

5 This apparent anachronistic presence of the family farm in Kansas is due to eight decades of statutes and constitutional mandates restricting non-local corporate takeover, protections put in place to defend a family ownership ethos that has been the norm in this region for over a century. Regarded as the backbone of rural Plains state opposition to corporate expansion during the Populist era of the late nineteenth century, family farm ownership emerged during World War I as the rhetorical locus of national ascent to world power, with wheat as the iconic "staff of life" that would feed the nation and mitigate world hunger. Though local family farms have incorporated, resistance to remote corporate takeover of the locally owned farm has been in place since the era of the first Great Depression, when locals were forced to fend off insurance companies that made land-title grabs at farms decimated in the Dust Bowl. See Bruce Johnson, "Corporate Farming," in *Encyclopedia of the Great Plains*, ed. David J. Wishart (Lincoln: University of Nebraska, 2011), http://plainshumanities.unl.edu/encyclopedia/doc/egp.ag.023. In early 2013, Kansas Secretary of Agriculture Dale Rodman called for repeal of these laws, commending corporations operating in other states for their performance

as *citizens*: "[T]hese industries are now modern, efficient and excellent corporate citizens." See Scott Rothschild, "Statehouse Live: State Agriculture Secretary Calls for Repeal of Laws Restricting Family Farms," *Lawrence Journal-World*, January 15, 2013, www2.ljworld.com/news/2013/jan/15/statehouse-live-state-agriculture-secretary-calls-/.

6 We share some of the goals and methods of the photographers and artists associated with the Center for Land Use Interpretation (www.clui.org/). However, whereas that organization's photographic work tends to focus on unpeopled landscapes, we also bring attention to workers and human–machine interaction on our documentation of sites.

7 Roland Barthes, "The Grain of the Voice," in *Image, Music, Text*, trans. Stephen Heath (New York: Hill and Wang, 1977), 180–183. These concepts are interestingly developed in Michael Szekely, "Gesture, Pulsion, Grain: Barthes' Musical Semiology," *Contemporary Aesthetics*, December 18, 2006, www.contempaesthetics.org/newvolume/pages/article.php?articleID=409#FN29link. See also Julia Kristeva, *Revolution in Poetic Language*, trans. Margaret Waller (New York: Columbia University Press, 1984), 87.

8 Associated Press, *Staff of Life Waivers under Weight of Humanity*, November 8, 2010, www.amigura.co.uk/news/science/climate_change/2-20-2010-11.html.

9 For documentation of the 2012 and 2013 droughts, see the United States Drought Monitor Map Archive at http://droughtmonitor.unl.edu/MapsAndData/MapArchive.aspx. The US Drought Monitor, established in 1999, is a weekly map of drought conditions produced jointly by the National Oceanic and Atmospheric Administration, the US Department of Agriculture, and the NDMC-UNL.

10 Walking down grain refers to the practice of employees walking on clogged grain in a storage facility to make it flow. Small farms are exempt from much agricultural labor regulation as well as laws pertaining to labor by children working on farms owned by their parents. Following an all-time high grain-entrapment death rate of 26 in 2010, the US Department of Labor proposed legislation prohibiting underage workers from entering silos, but it was withdrawn following widespread protest from farmers and politicians. On the described proposed legislation see US Department of Labor, "Labor Department Statement on Withdrawal of Proposed Rule Dealing with Children Who Work in Agricultural Vocations," press release, April 26, 2012, www.dol.gov/whd/media/press/whdpressVB3.asp?pressdoc=national/20120426.xml#.UJq3x_UYRXE.

11 Organized for the George Eastman House Museum of Photography in 1975 by assistant curator William Jenkins, the exhibition toured to Los Angeles and Princeton and was recreated as a traveling exhibition with two-thirds of its original works by the Center for Creative Photography and George Eastman House in 2009. See William Jenkins, *New Topographics: Photographs of a Man-Altered Landscape* (Catalog) (Rochester, N.Y.: International Museum of Photography at the George Eastman House, 1975). Also see Britt Salvesen with Alison Nordstrom, *New Topographics* (London: Steidl & Partners, 2010). See also Lisa Cartwright, "Topographies of Feeling: On Catherine Opie's American Football Landscapes," in *Feeling Photography*, eds. Elspeth Brown and Thy Phu (Durham, N.C.: Duke University Press, 2014).

12 Roland Barthes famously called the *studium* "a kind of education (knowledge of civility, 'politeness') that allows discovery of the operator," suggesting to me

that we should be attentive not to the intentionality of the photographer but to what we can sense of the intersubjective work between the photographer and her subject, whether the latter is a living being or a landscape. See Barthes, *Camera Lucida: Essays on Photography*, (New York: Hill and Wang, 1994), 28.

13 William Jenkins, *New Topographics*.

14 This phrase references the classic by Leo Marx, *The Machine in the Garden: Technology and the Pastoral Ideal in America* (New York: Oxford University Press, 1964).

Chapter 13

The face as a medium

Amit Pinchevski

If there is one thing that brings together the three themes of this collection – ethics, image, and technology – it is the face. As all the contributions to this volume masterfully demonstrate, the ultimate benefit in thinking across disciplines and paradigms is the opportunity to rethink concepts, especially those that appear to be self-evident or are taken for granted. Face is something that we all have yet is still unique to each and every one. It is the organ by which we best present and express ourselves and at the same time that which immediately betrays our origin, age, and mood. We are quick to extend our face to the world but as quick in wishing to hide or protect it from the world. We face each other regularly, but in facing we enact a singularly human combination of biology, culture, and morality. The face is our most basic medium of sociality. Like every medium, it is what stands in the middle, it both connects and separates – connects by virtue of separating – producing commonality while maintaining its in-betweenness. It is possible to identify three functions of the face as a medium: it is the surface by which we appear; the interface through which we interact; and the face-to-face in which we care for others. Regarding the face as a medium is meant here as a heuristic tool, with the aim of offering some reflections on the way technical media configure the social life of the face. "God has given you one face, and you make yourselves another," says Hamlet to Ophelia, a statement that might acquire new meanings in our age of Facebook, selfies, and facial-recognition technologies.

The great political theorist of the face is Hannah Arendt. In key sections of her work she sets up the face as a medium of appearance. The Anglo-French "visage" (countenance, portrait) is suggestive of that: its Latin origin *vis* (as in vis-à-vis) comes from *videre*, "to see" (the source "vision," "view," and "voyeur"). To appear in public is a fundamental political act, as Arendt argued influentially. Invoking the Athenian democratic prototype, Arendt reasserts the *polis* as the space of appearance, "the space where I appear to others as others appear to me, where men exist not merely like other living or inanimate things but make their appearance explicitly."[1] Appearance is not simply a matter of visibility. The space of appearance comes into being when people come together and face each by means of speech and action; it is a collaborative endeavor to maintain this space precisely as public space, prior to and outside of any institutional manifestation. Yet what was a distinctively face-to-face activity in ancient times takes a whole different turn in modern times when media do most of the appearing. This realization is developed most thoroughly by Roger Silverstone, who proposes the notion of the "mediapolis" as the contemporary incarnation of the space of appearance. "Contemporary media," he writes, "enable a face-to-faceness which, both in broadcast and interactive modes . . . involves the coming together of speech and action"; the media produce in "an intensely technologically mediated form, the discursive and judgmental space of the *polis*."[2]

Global in scope and cosmopolitan in nature, this mediated space of appearance nevertheless has the face-to-face as its organizing principle, and the face as its normative figure. Silverstone insists that despite the vast systems of sending and receiving, media production and consumption are still carried out facially: on the sending side, the interpersonal context of professionals (journalists, documentarians, producers); on the receiving side, the interpersonal context of audiences in their everyday lives; and in between, on the screens, the face reigns not only as a typical appearance but also as a typical mode of address that simulates face-to-face interaction. This is what Paddy Scannell calls the "for-anyone-as-someone structure"[3]: the curious yet utterly mundane fact that when I watch the news on television the anchorperson looks directly at me, addressing me in the second person – and this is true simultaneously to any and all other viewers. This mode of address is distinctive to broadcasting and has its roots in the early days of radio. On television, however, it is the face that pivots this mediating structure that both links and holds in tension private and public, personal and impersonal.

The obvious bears explication; never before have people seen so many of their own species as we and our most recent ancestors have since the advent of photography, film, and television. And our generation has probably seen more than all before us combined. The full significance of this historical development might still elude our understanding. Silverstone rightfully calls for considering the challenge this exponential visibility produced by modern media poses on our ethics and politics, especially when it comes to giving a face to the distant and the stranger. I will return to this later. Let us first consider a different challenge having to do with the way digital media reconfigure the visibility of the face.

Whereas analog visual media – be they photography, film, or television – reproduce images by means of chemical or electrical reaction to continuous variations of light, digital media sample these continuous variations and turn them into digital code: a series of zeros and ones. The image (or rather the numerical series representing it) is then available not only for reproduction and manipulation (already possible with analog media) but also for processing by algorithms that treat the image as pure data. Whereas analog media needed human intervention at some point in the production, reproduction, and evaluation processes, digital media can go through the process without consulting the human visual sensory. Digital technologies, to quote Friedrich Kittler, "are able to read and write by themselves."[4]

Consider facial-recognition technologies: they "see" the face as a particular configuration of surface measurements to be processed analytically and statistically. The principal users of these technologies are, of course, security and police agencies: a recently leaked report reveals that the American National Security Agency utilizes sophisticated programs that process millions of faces per day.[5] Garnered from emails, text messages, videoconferences, and social media, these images join an already vast government image database of drivers' licenses, visas, and border controls. Thus, surveillance is achieved not by means of controlling and limiting but rather by exploiting accessibility and visibility; surveillance grows in direct proportion to the laissez-faire of digital media. Indeed, our private digital devices such as cellphones, cameras, and computers are miniature face detectors, incorporating default face-recognition applications. Likewise are social media networks: Facebook employs a program called "DeepFace" in order to tag photos (based on a technology developed by the company Face.com, now owned by Facebook); online dating sites and applications employ the technology to predict possible matches ("send us a photo of your ex we'll find you the next"), but also to screen out known felons.[6] When we see a face on the screen, it has likely already been seen by eyeless algorithms. There are, to be sure, historical precedents: face-aggregating technology goes back to mid-nineteenth-century photography and to the likes of Duchenne de Boulogne, Sir Francis Galton, and Jean-Martin Charcot.[7] Yet the catalogs and composites they produced were still meant for human eyes – specifically, those of the position of privilege and power. Today's technologies of face aggregation work literally out of sight.

The implications of the above on Arendt's "space of appearance" are considerable. Appearance, always a contested political act, seems to have taken new dimensions, at once broadening and delimiting the political valence of the face. On the one hand, if we consider the mediapolis as covering both broadcasting and digital media then the potential to appear in public seems to be growing tremendously. This is not to say that the mediapolis is free of bottlenecks and gatekeepers, nor to suggest that all appearance is of importance, but rather to acknowledge the new affordances digital technology entail for appearance. On the other hand, with the exponentiation of digital surveillance, appearance is being subjected to new scopic regimes: for what it means for the face to appear in public is increasing exposure to surveilling violence while engaging in political action. This was evidenced by recent demonstrations under the Occupy movement: while clearly a battle for a space of appearance – and an exemplary case for the employment of social media to that end – there were those who demanded removing all cameras and some even opted to wear scarfs on their faces – precisely to thwart authorities' facial-recognition apparatus.[8] As the space of appearance becomes more heterogeneous, involving both human and nonhuman observers in an "actor–network" constellation, to use Bruno Latour's term, the political impetus behind appearance develops to include modes of visibility together with modes of invisibility, transparency, and stealth. Sometimes disappearance serves best the logic of the space of appearance.

The great sociologist of the face is Erving Goffman. Like Arendt, Goffman considered face and voice as intertwined and fundamental to social relation. Yet whereas for Arendt the face is a medium for political action – and as such at once private and public, personal and impersonal – for Goffman the face is a medium for impression management, a private medium under pubic scrutiny. To Goffman, social interaction is essentially twofaced: in the theater of the face-to-face all actors are required to play a role, negotiate backstage and frontstage behaviors, and recite the right script at the right scene. Here, too, etymology is suggestive: the Greek *prosopon* (literally "before the eyes") means both face and mask. The Latin equivalent *persona* (both person and mask) has a dubious yet beautiful source in *personare*, "sound through," as in the voice of an actor behind the mask. The words "face" and "fake" are likely cognates of the Latin *facere*, "to do" or "to make." Thus the face is a medium that one has to continually produce and reproduce in order to perform effectively in the presence of others. This is what Goffman calls "face-work": the constant effort to gain and maintain face in one's dealing with others, and to save face should one happen to lose it. "To study face-saving is to study the traffic rules of social interaction."[9]

Goffman was hailed fittingly as the "theorist of co-presence."[10] Yet similar to the way Plato's writing both confirms and undermines the teaching of Socrates on the primacy of speech, so Goffman's followers both affirm and betray his priority to the face-to-face. And, like in the case of Jacques Derrida's critique of logocentrism, at issue is the underlying inscription technology underwriting the face-to-face. This is because much of the work done in sociolinguistics – and indeed this field of study as a whole – would be inconceivable without a technology for inscribing and deferring speech: tape-recording. The fullness of human talk could become accessible only once captured on tape by nonhuman ears. Thus, Harvey Sacks's groundbreaking work in conversation analysis was based on recordings of a suicide hotline, as well as on recordings of his own lectures (thus making himself both examiner and examinee of conversation analysis). Likewise the work of his followers, Emanuel Schegloff and Gail Jefferson, on turn-taking, micro gestures, and laughter could not been achieved without time–axis manipulation of audio recording.[11] While people have been talking for millennia, the study of conversation could not have emerged without the attendant cultural technique: "media that process the observation, displacement and differentiation of distinctions."[12] The inner workings of social co-presence become discernible only through technology of archiving presence.

"Face-work" is nevertheless an eminently palpable preoccupation for anyone who is in the presence of others. It stands to reason that technologically mediated communication would elevate some of the burden of co-presence. In the mid-1990s, when internet access was becoming popular, there were fierce debates about online interaction carried out in IRCs (Internet Relay Chats, or chat-rooms) and MUDs (Multi-User Domains). There were those who celebrated the liberating potential in online chatting, in pretend, and role plays, precisely because faceless and hence unhindered by origin, color, age, or gender – a virtual public sphere. As Sherry Turkle, one of the key proponents of this view, writes: "As players participate, they become authors not only of text but of themselves, constructing new selves through social interaction."[13] The release from co-presence, visibility, and "face-work" was said to promote freedom of expression and license for experimentation. And then there were those who regarded this release as destructive: anonymity and distance were said to produce deception, intemperance, and alienation. In faceless interaction there is no face to lose, nor one to save. Patrick Marber's 1997 play (and later film) *Closer* illustrates this bleak view: in a memorable scene the two male protagonists interact in an online sex chat while one identifies himself as Anna (a character with whom he is in love), then proceeds to play a practical joke on the other by arranging to meet in real life; this scene is the turning point after which everything starts to crumble. At stake is truth, or rather, the facility of dishonesty. This admittedly sketchy description nevertheless brings into focus the hopes and fears of the early years of the internet in leading faceless existence online.

It is striking how far removed we are now from that debate. Today's social media are rampant with faces, Facebook being the paradigm, literally.[14] Life on the screen nowadays is pointless without propagating face and identity. Mark Zuckerberg's ultimate achievement was convincing everyone that anonymity isn't fun anymore: "[W]hen Facebook was getting started, most people didn't want to put up any information about themselves on the Internet . . . So, we got people through this really big hurdle of wanting to put up their full name, or real picture."[15] The convergence of online platforms and mobile media has given rise to a culture of sharing visibility. What better typifies this trend than the 2013 Oxford Dictionary word of the year, "selfie": "A photographic self-portrait; *esp.* one taken with a smartphone or webcam and shared via social media." If terms such as cyberspace and virtual reality now seem outdated it is because they refer to a time when the internet was a space for an alternative existence, removed from one's appearance and identify. Multi-user interaction has given way to tweeting, posting, online chatting and mini-broadcasting. Rather than surrogate sociality, social media are the continuation and expansion of social life by other means. In this respect, Facebook calls for a revision of "face-work": a collectively mediated impression management by means of "likes," "comments," "tags" and "shares" – the traffic rules of social media.

The great philosopher of the face is Emmanuel Levinas. An orthodox Jew, Levinas unapologetically strived to instill the morality of the prophets into the Western tradition. To him, the face is a medium of ethics: the interface by which one encounters another as Other. The face of the Other exceeds the idea of the Other in me. Neither a theme under one's gaze nor a combination of features forming an image, the face resists full visibility: "The face of the Other at each moment destroys and overflows the plastic image it leaves me."[16] How, then, does the face transcend its own and any image? By speaking, addressing: "[T]he face speaks to me and thereby invites me to a relation incommensurate with a powers exercised, be it enjoyment of knowledge . . . To speak to me is at each moment to surmount what is necessarily plastic in manifestation."[17] This duality is encompassed by the Hebrew *panim* (face): the plural of *pan* (aspect, facet), it shares a common root with *pniya*, "to turn," "to address." The face faces: again face and voice are connected; yet in Levinas's case it is by way of undoing the ostensible. So for the face to maintain its ethical import it must somehow appear beyond what is seen, beyond representation; and in order to do so, it must somehow issue a call, an expression that fractures the *sur*face.

Levinas left little guidance as to how to apply his ideas, perhaps rightly so. Indeed, application, in the strict sense, would immediately pose a threat to the irreducibility announced by the face. One way to proceed under his inspiration is by attending to the necessary paradox integrated in the face. When it comes to the context of media, two notable examples come to mind. As mentioned before, Silverstone's account focuses on the ethical challenge in representing the Other's face in the media. The concept he develops to that end, following Levinas, is "proper distance": the degree of "proximity required within our mediated interrelationships if we are to create and sustain a sense of the other sufficient not just for reciprocity but for a duty of care, obligation and responsibility."[18] Such is the challenge of being close but not too close as to assimilate, of being distant but too distant as to be indifferent. Silverstone's idea of the mediapolis can be understood as bringing Arendt's space of appearance together with Levinas's visibility–invisibility structure of the face. Representations are what the media produce, and it is by working constructively against the grain of representation, as per Silverstone, that we may commit ourselves to the ethical challenges the media present us with.

The other notable example is Judith Butler's discussion of the face as the epitome of precarious life. Further to Levinas's assertion of the face as the origin of speech, Butler explores the conditions by which the face can maintain what she calls "the structure of address," that is, the situation in which we find ourselves addressed in ways that we cannot escape or ignore.[19] The commonsense assumption is that granting visibility to the face is humanizing while denying or effacing the face is dehumanizing. While the latter is certainly true, Butler argues counterintuitively that sometimes personification can be equally dehumanizing. This is evident in the way the media produce the face as an image of evil (the faces of Bin Laden and Arafat, for instance), but to the same measure as an image of freedom, as in a picture of Afghan girls stripping off their burkas. In both cases the face is placarded and is thus deprived of its irreducibility as a face. In this respect, faces produced as evil or good are figures usurped for ideological interpellation, which of course presupposes an underlying structure of address to be usurped: this face speaks, and the way it speaks stands for evil or good. Butler contends that for an image to convey the face it must never succeed in capturing the face: "There is something unrepresentable that we nevertheless seek to represent, and that paradox must be retained in the representation we give."[20] Both Silverstone and Butler are invested in keeping Levinas's paradox alive through the question of representation, which, although central, might also be limiting, as I suggest elsewhere.[21]

And then there is the mundane context in which media regularly and ubiquitously circulate images. We live in a time where we look at numerous faces every day without them being able to look back – a fact that astonishes us less than it should. As Paul Frosh notes, the headshot has been a dominant pictorial figure of television throughout its history, surviving various conventional and technological transformations. Of all other things, television is a veritable face machine. This seemingly trivial point harbors nascent morality: allowing the faces of strangers to appear in the intimacy of our homes, without them causing threat or alarm, serves to promote a "desirable cosmopolitanization of the home." Moreover, the accumulative presence of these faces amounts to moral significance insofar as it produces "a *serial aggregate* of human similarity and connectedness, an interminably fluctuating and ever-present composite image: the face of humanity itself."[22] What is compelling about this argument is that it brings media to bear on the conditions by which we come to relate to others, both near and far. The problem, however, lies in its conception of the face as exclusively surface, completely bereft of its twin facet of address. A morality based on the residual impression of faces would lead at most to tolerance, but not to responsibility. That images cannot look back at us does not mean they cannot address us, particularly when expressing vulnerability and suffering – a theme that preoccupied the late writings on photography of both Roland Barthes and Susan Sontag.[23] The challenge remains of developing new thinking of the mediated face as a medium of address together with and despite being a medium of appearance.

If there is any benefit in this exercise of rethinking the face through the logic of media, it is in exploring new ways of how ethics and media might implicate each other. Which brings me to one final speculation: media have become a factor in ethics, not merely a case for ethics. This is true specifically to the relation between the particular and the universal. Through the face, the media now give new concreteness – and urgency – to that age-old question. Kant developed his formula of the universal law without ever leaving the vicinity of his hometown Konigsberg. Today's philosophers cannot indulge in such abstracted universalism; for they, like most of us, live in a world where the universal has a face, not just one but many, and the particular can be pluralized far and wide. Media are what stand for "relation" between the particular and the universal, and, as such, are the a priori context of any present and future ethics. Media are not merely functional in-betweens but the substance from which ethical coordinates arise. So, perhaps instead of speaking of distinct categories (and the relation linking them), the particular and the universal might be thought of as asymptotes along a continuum: diametrically opposed, they converge at some infinite imaginary point. Ethics are profoundly in media res.

Notes

1 Hannah Arendt, *The Human Condition* (Chicago: The University of Chicago Press, 1958), 198–199.
2 Roger Silverstone, *Media and Morality: On the Rise of the Mediapolis* (London: Polity, 2007), 29–30.
3 Paddy Scannell, "For-anyone-as-someone Structures," *Media, Culture and Society* 22, no. 5 (2000): 5–24.
4 Friedrich A. Kittler, *Literature, Media, Information Systems*, ed. John Johnston (Amsterdam: G+B Arts, 1997), 147.
5 James Risen and Laura Poitras, "N.S.A. Collecting Millions of Faces from Web Images," *The New York Times*, May 31, 2014, www.nytimes.com/2014/06/01/us/nsa-collecting-millions-of-faces-from-web-images.html?_r=0.
6 Dino Grandoni, "Facebook's New 'DeepFace' Program Is Just as Creepy as It Sounds," *The Huffington Post*, March 18, 2014, www.huffingtonpost.com/2014/03/18/facebook-deepface-facial-recognition_n_4985925.html; James Vincent, "NameTag: Facial Recognition App Scans Faces for Dating Profiles, Criminal Background," *The Independent*, January 9, 2014, www.independent.co.uk/life-style/gadgets-and-tech/facial-recognition-app-scans-strangers-faces-for-dating-profiles-criminal-background-9049568.html.
7 See discussions in Philip Prodger, "Illustration as Strategy in Charles Darwin's *The Expressions of the Emotions in Man and Animals*," in *Inscribing Science: Scientific Texts and the Materiality of Communication*, ed. Editor Timothy Lenoir (Stanford, Calif.: Standford University Press, 1998), 140–181; George Didi Huberman, *Invention of Hysteria: Charcot and the Photographic Iconography of the Salpètrière* (Cambridge, Mass.: MIT Press, 2004); Sharrona Pearl, *About Faces: Physiognomy in Nineteenth-Century Britain* (Cambridge, Mass.: Harvard University Press, 2010).
8 Nathan Jurgenson, "Occupy: Anonymity or Transparency?" *The Atlantic*, December 15, 2011, www.theatlantic.com/technology/archive/2011/12/occupy-anonymity-or-transparency/250053/.
9 Erving Goffman, *Interaction Ritual: On Face-to-face Behavior* (New York: Doubleday, 1967), 12.
10 Anthony Giddens, *Social Theory and Modern Sociology* (Stanford, Calif.: Stanford University Press, 1987), 115.

11 See Harvey Sacks, *Lectures on Conversation, Volumes 1–2*, ed. Gail Jefferson and Emmanuel A. Schegloff (London: Wiley-Balckwell, 1995).
12 Bernhard Siegert, "Cacography or Communication? Cultural Techniques in German Media Studies," *Grey Room* 29 (2001): 26–47.
13 Sherry Turkle, *Life on the Screen: Identity in the Age of the Internet* (New York: Simon & Schuster, 1995), 12.
14 For a most literal illustration, see http://app.thefacesoffacebook.com. I thank Yuval Dror for this and other Facebook facts.
15 Michael Zimmer, "Zuckerberg's Theory of Privacy," *The Washington Post*, February 3, 2013, www.washingtonpost.com/lifestyle/style/mark-zuckerbergs-theory-of-privacy/2014/02/03/2c1d780a-8cea-11e3-95dd-36ff657a4dae_story.html.
16 Emmanuel Levinas, *Totality and Infinity: An Essay on Exteriority*, trans. Alphonso Lingis (Pittsburgh, Pa.: Duquesne University Press, 1998), 50–51.
17 Levinas, *Totality and Infinity*, 198, 200.
18 Silverstone, *Media and Morality*, 47.
19 Judith Butler, *Precarious Life: The Powers of Mourning and Violence* (New York: Verso, 2004), 129.
20 Butler, *Precarious Life*, 144.
21 Amit Pinchevski, "Levinas as a Media Theorist: Toward an Ethics of Mediation," *Philosophy and Rhetoric* 47, no. 1 (2014): 48–72.
22 Paul Frosh, "The Face of Television," *Annals of the American Academy of Political and Social Science* 625 (2009): 90, 93–94.
23 Roland Barthes, *Camera Lucida*, trans. Richard Howard (New York: Hill and Wang, 1982); Susan Sontag, *Regarding the Pain of Others* (New York: Farrar, Straus, and Giroux, 2002).

Index